DUST AND FUME CC
A USER GUIDE

Revised Second Edition

D.M. Muir (Editor)

D.A. Jones, HSE (revisions of second edition)
J.O'Hara, Denton Hall Burgin and Warrens and
J.L. Barnett, HMIP (additional contributions)

INSTITUTION OF CHEMICAL ENGINEERS

Distributed exclusively in the USA and Canada
by VCH Publishers, Inc, New York

The information in this guide is given in good
faith and belief in its accuracy, but does not
imply the acceptance of any legal liability or
responsibility whatsoever, by the Institution, by
the editor or by individual members of the
Working Party for the consequences of its use or
misuse in any particular circumstances.

Published by
Institution of Chemical Engineers,
Davis Building,
165–171 Railway Terrace,
Rugby, Warwickshire CV21 3HQ, UK.

Distributed exclusively in the USA and
Canada by
VCH Publishers, Inc,
220 East 23rd Street,
Suite 909,
New York,
NY 10016–4606,
USA.

ISBN 0 85295 287 2

First edition 1981
Second edition 1985
Reprinted 1987, 1988 and 1991
Revised second edition 1992

Printed in England by BPCC Wheatons Ltd, Exeter.

WORKING PARTY MEMBERS

The first edition of this book was produced by the Engineering Practice Committee Working Party on Dust and Fume Control (1981) with the following membership:

C.R. Smith	Peabody Holmes Limited
R.W.K. Allen	AERE Harwell
M.J. Ashley	Lodge Cottrell Limited
R. Harrison	ICI Limited
B.J. Squires	Tilghman Wheelabrator Limited
G.L. Sneddon	Mikropul Ducon Limited
J.I.T. Stenhouse	Loughborough University
A. Dick	James Howden Limited
J. Bryden	Institution of Chemical Engineers

Special contribution to first edition
Industrial Gas Cleaning Association

CONTENTS

1. INTRODUCTION

This guide provides information for the selection, installation and operation of gas-cleaning equipment associated with dusts and fumes. For the present purpose, fumes are defined as particles of very small (sub-micron) size. The emphasis is on solid particles, but the removal of liquid droplets is covered in that demisters are an intrinsic part of wet particle collectors. Throughout the text the use of the words 'dust', 'fume' or 'particle' have been selected as being appropriate in that context; the use of any one description should not be taken to be exclusive. Direct consideration has not been given to the treatment of odours or other gaseous pollutants.

The aim has been to provide a document which gives a non-specialist engineer advice on how to set about solving a particle emission problem either on an existing plant or for a new project. Detailed design matters on individual dust collectors have purposely been limited as these are adequately covered in other publications. Rather, the guide tries to give an overview of the subject as this was felt to be of greater value to most users. Throughout, the emphasis is on a total systems approach to the solution of dust and fume control problems. For the user requiring a more in-depth knowledge of the subject, the guide should be read in conjunction with the references.

Careful consideration has been given to the layout of this guide. It is suggested that prospective users should start by reading Chapters 2 to 7. The first requirement is to define why a dust collector may be needed and to quantify the extent of the problem. The user must also determine the standard of dust control required in order to satisfy any legal or other obligations. There is then the question of how effectively any dust released from the process can be captured, and what effect this will have on the inlet gas volume and solids concentration entering the collector.

The heart of this guide is the selection procedure given in Chapter 7. This should enable a prospective user to make an initial technical choice between the five main types of dust collector. Having selected one or more possible collector types, the reader can then confine his attention to the appropriate

1

sections from Chapters 8 to 12. When the user has a clearer view of the plant which is appropriate technically, he should then continue with the rest of the guide in order to obtain a more complete understanding of all the factors to be considered. Particular stress has been put on the role of the fan and its drive, and the need to carefully match the fan to the dust collection system.

The guide is not intended to be an exhaustive treatment of the subject of gas cleaning. It is hoped, however, that by covering those points relevant to the selection, installation and operation of the main types of gas cleaning equipment, it provides a useful design guide for the practising engineer.

2. THE SOLUTION OF DUST CONTROL PROBLEMS

The purpose of a dust control system is to capture, collect and deliver in a convenient form particulate material which would otherwise disperse into the general atmosphere, causing a nuisance, a hazard or both. The types of equipment used in dust control are also employed for gas-solids separation on operations where particulate matter is being processed in bulk for further use, sale or disposal.

Selecting a successful combination of equipment will depend on many factors involving a blend of technical, economic, legal, geographic and management knowledge. Often no perfect solution exists and engineers will have to accept certain limitations which may affect plant sizing and operation. At the end of the day, however, the dust collection system has to perform satisfactorily. Failure to meet objectives can usually be traced back to inadequacies in the original definition of the problem so extra care at this stage will provide a sound basis from which to develop the solution.

2.1 THE VARIOUS ASPECTS OF DUST CONTROL

A systematic approach to the solution of a dust control problem will essentially involve consideration of a number of different aspects of the subject:

(1) IDENTIFICATION OF THE EXACT NATURE OF THE PROBLEM

A process may produce dust but not in sufficient quantities for the company to consider it to be a problem. However, pressure to improve the situation may be applied by Her Majesty's Inspectorate of Pollution (HMIP), by the local authority, by the Health and Safety Executive (HSE), by the process operators and/or their Unions, by neighbouring companies or by local residents, by the need to improve housekeeping or reduce the loss of valuable product or by the company's social conscience.

It is, of course, important to know whether the dust is 'inert and of nuisance value only' or whether it is toxic, flammable or explosive. The user's obligations under the law are dealt with in Chapter 3.

(2) PREVENTION OF THE DUST BECOMING A NUISANCE

If it is possible to prevent the dust becoming a nuisance, the problem is essentially solved. The various methods of preventing the dust becoming airborne are outlined in Chapter 4. Even if only partially successful, the techniques used will minimise the quantity of dust to be controlled and hence reduce the ultimate cost of environmental control equipment.

(3) CONTAINMENT AND CAPTURE OF THE DUST

If airborne dust is unavoidable, there are essentially three methods available for its containment and capture: total enclosure, partial enclosure and hooding. With the last two methods, a specific volume of air will have to be extracted to prevent egress of dust from the process. Methods of containment are dealt with in Chapter 5.

(4) TRANSFER OF THE DUSTY GAS OR AIR IN A DUCT SYSTEM

Proper design of the ducting is essential for the successful operation of any dust control plant. This is one aspect of the subject which frequently receives insufficient attention and is often the cause of maloperation of the plant. Some prior consideration of the ducting layout should always be made before siting the main items of equipment. Where several ducts feed a main duct, the duct system has to be correctly 'balanced' to achieve the correct gas or air flowrate from every extraction point. The principles of ducting design are considered in Chapter 6.

(5) GAS COOLING OR CONDITIONING PRIOR TO DUST COLLECTION

For a variety of reasons, dust-laden gases are often cooled prior to collection. These may be to recover heat, to reduce the volume of gas to be handled or to protect filter fabrics or corrosion-protective linings against excessive temperatures. The different methods of gas cooling are dealt with in Chapter 7.

Conditioning, on the other hand, might involve increasing the humidity of a dry gas stream to assist precipitation of a difficult dust or adding a second particulate material, such as lime, to the gas stream to facilitate fabric cleaning in a baghouse.

(6) SEPARATION OF THE DUST FROM THE AIR OR GAS STREAM

There is a wide choice of equipment used in dust control: cyclones and inertial separators, wet dedusters and demisters, electrostatic precipitators, fabric filters

4

and deep bed filters. A proper technical evaluation of the type of equipment best suited to a particular application will depend upon many factors. These will include process considerations, operating conditions, dust characteristics, required performance, site layout, etc. A procedure for plant selection is presented in Chapter 7 along with the type of information normally required to make the initial choice. The different types of collectors are considered in more detail in Chapters 8 to 12. 7- 10 not assessable

(7) HANDLING AND DISPOSAL OF COLLECTED DUST AND/OR SLURRY
It should always be remembered that the problem is not solved once the dust has been collected. The dust may be collected dry, in which case there is a potential secondary dust problem, or it may be collected as a slurry. The type of further treatment required will depend on whether the dust is a product, whether it can be recycled to the process or whether it is for disposal. Guidance on the handling and disposal of dusts and slurries is given in Chapter 13.

(8) AIR-MOVING MACHINE OR FAN
Fan selection is a complex business and advice should be sought from the experts. The successful performance of dust control plant is critically dependent upon the correct selection of the fan. It should be noted that the handling of dust-laden gases represents one of the most difficult and demanding applications in the field of fan engineering. A compromise has frequently to be made between the design of fan impeller that makes the most efficient use of power and one that is unlikely to go out of balance because of dust deposition on the blades. The different types of fans used for gas cleaning and their chief characteristics are dealt with in Chapter 14. The problem of fan noise is also included.

(9) CONTROL AND INSTRUMENTATION
This can vary from a simple manometer to indicate the pressure drop across a filter or scrubber to a fully-integrated control system to automate the complete process and protect the plant against high temperatures, dewpoint conditions, leakage, etc. Guidance on this subject is given in Chapter 15.

(10) DISPERSAL OF THE EXHAUST GASES
Assuming that the arrestment plant has been correctly designed to achieve an acceptable concentration of particulate matter in the exhaust gases, it is important that the stack is suitably located and is of a sufficient height to ensure that

5

any residual pollution is adequately dispersed to the atmosphere. The dispersal of exhaust gases is the subject of Chapter 16.

In addition to the above aspects of dust control, Chapter 17 deals with the economic evaluation of the plant and Chapter 18 with methods of testing dust-laden gases which are currently used both for obtaining data prior to design and for testing plant performance after installation.

2.2   NEED FOR A SYSTEMATIC APPROACH TO THE PROBLEM

The order in which the different aspects of the problem are tackled is not necessarily the same as the sequence (1) to (10) outlined above; both the order and relative importance of the different aspects will vary from problem to problem, although the approach will almost certainly begin with the identification of the problem. What is important is that the engineer who is responsible adopts a systematic approach to the solution of the problem in which all the important issues are considered and dealt with in the most appropriate order. One of the main objectives of this guide is to encourage a disciplined approach to the design of dust control systems. A good example of a systematic procedure for the control of dust, which should not be regarded as applicable to all dust control problems, is given in Reference 27.

2.3   OBTAINING HELP AND ADVICE

There are a number of sources from which both skilled and inexperienced users can get help in formulating a specification of their requirements. These may include:

* the engineers who designed and operate the main process from which the dust emanates;
* HMIP;
* local authorities (Environmental Health Departments);
* HSE
* reputable dust collector contractors, preferably those with a wide range of equipment and relevant experience;
* industrial or academic consultants;
* trade and research organisations;
* other users of similar processes.

However the specification is achieved, it is essential that a potential user of dust collection equipment provides accurate and full data to the equipment suppliers to ensure that the equipment offered to him is fully in line with his requirements. It is also in the user's interest to disclose any unknowns or areas of uncertainty rather than to include excessive safety margins.

It is important to note that in the dust control industry it is common practice for contractors to offer a complete design package rather than to sell individual collectors, except perhaps for the simpler applications. Contractors frequently insist on designing the complete plant as a condition in honouring equipment performance guarantees.

3. ENVIRONMENTAL, HEALTH AND SAFETY REQUIREMENTS

The Health and Safety at Work etc Act (1974) imposes upon employers a duty to provide a safe place of work and a safe method of working for their employees. It also imposes upon designers, manufacturers, importers or suppliers duties with respect to the safety of any article for use at work. In addition, there are requirements covering the health and safety of persons not directly employed. Atmospheric emissions from the plant will be subject to control either by the local District Industrial Pollution Inspector for processes registered under the Alkali Act or the local authority Environmental Health Department. Controls for registerable processes extend to prior approval of design features relevant to air pollution, including the provision of gas cleaning equipment that may be necessary to prevent or minimise the emission.

With regard to process discharges, it is important to record that the Environmental Protection Bill received Royal Assent in 1990. The enactment of this Bill introduced a number of fundamental changes compared with previous industrial pollution control legislation. All new processes and substantially modified existing processes currently registered with HMIP and/or defined within Part A of Schedule 1 of the Environmental Protection (Prescribed Processes and Substances) Regulations 1991 will, henceforward, be subject to regulation by HMIP in regard to all process-related solid and liquid discharges as well as discharges to atmosphere. Existing processes currently registered with HMIP in accordance with regulations made under the provisions of the Health and Safety at Work etc Act 1974 will be progressively brought within the legislative ambit of the Environmental Protection Act 1990, over the period extending to January 1996. Processes defined within Part B of Schedule 1 of those same regulations will be subject to regulation by the local authority Environmental Health Department but only in regard to discharges to atmosphere.

The previous requirement for an operator to secure prior approval has been replaced by a more formal procedure involving an application to the appropriate regulating authority. That procedure is much akin to existing planning procedures; a notice of the application has to be placed in a local paper, interested members of the public should have access to the application and be able to take copies of the application, and the inspector should take any subsequent comments made by members of the public in regard to that application into account when determining the application. The application to HMIP or the local authority should be accompanied by the appropriate fee and the information supplied with the application should provide adequate detail in regard to the process, the control provisions and implementation details in regard to the operation of the process.

The corner stones to integrated pollution control are those of the operator demonstrating within the application that the chosen process represents the Best Practicable Environmental Option (BPEO), that the control and operating provisions for that process will meet the requirement of adopting Best Available Techniques Not Entailing Excessive Cost (BATNEEC), and that the impact of the process on the environment will be acceptable. In the case of a successful determination of an application, the authorisation document issued by the regulating authority will detail operational constraints and limitations applying to the satisfactory operation of that process. These limitations and constraints are subject to formal review by the regulating authority every four years. If it is considered appropriate, that review process may modify previous requirements to better reflect current knowledge of the effects of pollutants, improvements in control technology, etc.

Guidance on the procedures appropriate to making an application for authorisation of a process scheduled under the Environmental Protection (Prescribed Processes and Substances) Regulations 1991 may be obtained from HMIP or the local authority Environmental Health Department.

A number of smaller processes will continue to fall outside the scope of the Environmental Protection (Prescribed Processes and Substances) Regulations 1991. Such processes may be subject to control under the provisions of the 1956/68 Clean Air Acts or the Control of Pollution Act 1974. In such cases, advice should be sought from the local authority Environmental Health Department.

The Control of Substances Hazardous to Health Regulations (COSHH)[3] require that any circumstances where substances may be released are identified, their potential for harm assessed and the control measures to limit that potential identified. It is therefore important to know whether the dust is more than ordinary nuisance dust and, therefore, what Exposure Limit or other hygiene standard should apply. Occupational Exposure Limits are listed by the Health and Safety Executive[5] (HSE) and by the American Conference of Governmental Industrial Hygienists[15] (ACGIH).

Noise emitted by the gas cleaning equipment is also subject to control by the local authority under the provisions of the Environmental Protection Act. This will be particularly relevant if the installation is close to residential properties.

3.1 FIRE AND EXPLOSION

Many dusts, particularly those of organic origin, are capable of forming flammable dust clouds. If such clouds ignite when they are dispersed in air, a violent explosion may occur.

Section 31 of the Factories Act requires that all practicable precautions be taken to guard against dust explosions. The basis of such safety from explosions may be either:

(i) explosion prevention, or

(ii) acceptance of the possibility of an explosion and provision of a method of protecting personnel and equipment from its consequences.

Explosion prevention may be based on provision of an inert atmosphere, avoiding the formation of a flammable dust cloud or ensuring that no ignition sources exist. Explosion protection may be based on relieving, suppressing or containing the potential explosion.

Users of dust control equipment are strongly urged to refer to detailed information on the fire and explosion hazards of particulate materials and the associated safety precautions.

Further guidance on this subject is given in References 1, 2 and 7–11.

3.2 NOISE

In dealing with noise produced by industry, the subject can be divided into two areas: inside noise, which affects personnel employed inside a works' premises, and outside noise which affects personnel living in the vicinity of a works.

INSIDE NOISE

By virtue of the 1974 Health and Safety at Work etc Act, employers are required to do all that is reasonably practicable to protect the health and safety of persons at work. In interpreting these requirements, Factory Inspectors use the Noise at Work Regulations 1989 and the Guidance on these regulations[4]. This may be accomplished by virtue of machine design, enclosure or the fitting of silencers. Any legal requirements should be regarded as the minimum standard and in some cases these may not be sufficient to satisfy plant operators.

OUTSIDE NOISE

Noise which adversely affects the area surrounding works' premises in which it arises may be termed neighbourhood noise. The Control of Pollution Act 1974 covers such a noise nuisance. An individual aggrieved by noise nuisance has a choice of several courses of action. He may (a) complain to the Environmental Health Department of a local authority; (b) complain directly to a Magistrates' Court; (c) take civil action for noise nuisance at common law. The defence of 'best practicable means' is available for actions (a) and (b) but not for (c).

3.3   FACILITIES FOR TESTING

Where the final emission is to atmosphere, whether via a stack or not, test points should be provided in accordance with BS 3405: 1971, together with the necessary platform, handrails and access ladder. This facility should be regarded as an integral part of the plant design and not treated merely as an afterthought.

3.4   DISPOSAL OF SOLID AND LIQUID WASTES

Processes defined within Part A of the Environmental Protection (Prescribed Processes and Substances) Regulations 1991 will be subject to integrated pollution control. HMIP will wish to examine using detail supplied as part of the application made to HMIP for authorisation of the process, all matters relevant to the nature of such wastes, to the minimisation of such wastes and to the satisfactory disposal of such wastes. In the case of other processes, it should always be remembered that a problem that begins as a potential air pollution problem may end up as a liquid pollution and/or solids disposal problem. It follows, therefore, that the legislation governing the disposal of solids and liquid wastes should also be consulted at the identification of the problem stage. Further advice is given in Chapter 13.

not assessable but interesting

4. PREVENTION OF DUST BECOMING A PROBLEM

Before considering how a dust problem may be solved by the installation of equipment such as hoods, ducting, dust collectors, etc, it makes sound economic sense to examine the process to see if the dust nuisance can be avoided. The problem should be approached methodically, examining a sequence of alternatives against the process constraints. Even if one is not completely successful at least one has gained the satisfaction of minimising the dust to be captured and possibly the quantity of valuable product lost.

4.1 SUBSTITUTION

This is the first approach. The process and its economics should be examined to see if a non-toxic, a less toxic or a less dusty substance can be substituted for the material from which the dust derives. It may be possible to improve the situation sufficiently to make further action unnecessary. Examples of this technique are the replacement of asbestos lagging with mineral wools, urethane foam, etc and the substitution of aluminium oxide abrasives and calcium carbonate fillers for siliceous materials.

4.2 FINES

Is it necessary for the material to have fines small enough to become airborne? If not, can the material be obtained or manufactured without fines or with less fines or with fines in the form of agglomerates? Can additives be used to prevent dust becoming airborne?

Will the process accept the material as a solution or a slurry? Is it possible to produce the product in the form of pastilles, flakes, prills or granules?

Examples of this approach are the wide range of materials which are spray- or fluidised-bed-dried to give a granular product. Most fertilisers are prilled or granulated these days. Many pigments are treated with minute quantities of liquid dedusting agents. Another additive is long-chain PTFE polymer in powder form which is reported to bind the particles together and prevent dust without, say, destroying the dispersion properties of a pigment.

While considering the above techniques it should be remembered that the presence of fines in a product not only makes it dusty but generally hinders its flow properties. Granulation, etc will normally improve the flow properties, but liquid additives are likely to make the materials handling more difficult.

4.3   GENERATION OF FINES

Clearly, it is normally detrimental to the product to generate unnecessary fines in the equipment through which it passes. A study of the equipment upstream of the source of dust should be made to minimise, where possible, situations where fines are likely to be generated. With a raw material, much of this equipment may be on another site. Look for places where powder is moved at high velocity or where there are moving parts with close clearances, particularly rolling and rubbing motions.

Much damage can be done to the particles by such machines as dilute-phase pneumatic conveyors, cyclones, high-speed mixers, screw conveyors and elevators.

Some physical forms of powder are more prone to dust generation than others. Sharp crystals may lose their corners, particularly dendritic and needle crystals. Agglomerates and granules may be fragile enough to break down into their primary particles.

4.4   AIRBORNE DUST

If it is unavoidable to have fine particles in the product, it is sensible to keep to a minimum the opportunities for them to become airborne. The following conditions represent the most likely sources of airborne dust:

- air being passed through the material;
- the material being allowed to fall freely through the air;
- the material being agitated or tumbled;
- condensation of a sublimate or solidification of molten droplets;
- evaporation of solvent from a droplet of a solution;
- evaporation of liquid from a droplet of a suspension;
- ash or fume from a combustion or other high temperature process.

Attention to detail, particularly during the definition and design stages of new plant, can do much to reduce the problems of dust control. Not infrequently a dust problem can be solved without expenditure of capital and without the additional running and maintenance costs of dust collectors and fans.

4.5 USE OF SPRAY NOZZLES

This techique is generally applied to the processing of coarse solids which have associated fines that are likely to become airborne. It can be used to perform three basic functions:

• Pretreatment — Wetting the product material to cause agglomeration of fines prior to the dust-creating operation.

• Confinement — Isolation of the dust emission area by a thin curtain of spray in order to keep the dust within the treatment zone.

• Suppression — Capture of newly created fines by direct impaction with fine spray droplets, thereby returning the fines on to the mass of the product material.

Spray nozzles can be used on crushers, screens, conveyors, etc and are particularly useful for dust suppression on stockpiles. The success of the technique lies not only in the correct location of the spray nozzles but also on the correct choice of nozzle that will give the type of spray pattern best suited to the dust conditions at each treatment point. Regardless of the number of treatment points, the total moisture added to the solids is seldom more than 1% by weight. It should be remembered when deciding on the position of spray nozzles that access will be required for regular maintenance.

5. DUST CONTAINMENT AND CAPTURE

If, after examining the problem in the manner suggested in Chapter 4, it is still clear that airborne dust is inevitable, then total enclosure, partial enclosure or hooding will be required. Many operations that generate dust are enclosed, or at least partially enclosed, often for mechanical safety reasons as much as for dust control. Mills, crushers, elevators and some types of conveyors are typical examples. The more close and complete the enclosure around the point of dust generation, the more efficient will normally be the capture of dust and the lower the volume of extraction air required.

There are, however, practical and sometimes cost limitations in achieving this ideal. The need for safety, operator access, observations and maintenance, plus the proximity of other process components must be taken into account. In many situations, therefore, a hood may have to be used instead of an enclosure. The purpose of a hood is to capture the airborne dust as effectively as possible, consistent with inducing into the hood the minimum volume of uncontaminated surrounding air.

There are basically two different types of hoods, captor hoods and receptor hoods. These hoods may look similar, but their methods of operation are different and they must be designed on different principles. A captor hood is required to collect dust which would not otherwise enter the hood. Its required 'capture' velocity is the air velocity in front of the hood which will overcome the forces acting on the dust cloud causing it to change its direction of dispersal and flow into the hood. A receptor hood, on the other hand, must capture the dust and fume that is forced towards it by some means or other, such as by air displacement or by convection. In designing a receptor hood, it is necessary to know the volume and velocity of approach of the contaminated air.

5.1 TOTAL ENCLOSURE

Total enclosure means gas-tight enclosure. If the enclosure is allowed to breathe, even through such leaks as are commonly found in large diameter light flanges, then egress of dust will also occur. In the design of such an enclosure it is important to ensure that it will remain gas tight during its likely lifetime and after several openings and closings to maintain the equipment inside.

When considering total enclosure as a solution to a dust problem there are two potential disadvantages which should not be overlooked. The first is that an enclosed space is likely to have no movement of air within it or at best the air will move at a very low velocity which will not be uniform. This means that dust created within the enclosure will settle. Provision must therefore be made for the dust to return to the process. This can be achieved by making the enclosure part of the process equipment or perhaps providing it with a hopper bottom. Alternatively, it must be accepted that the enclosure will need to be opened periodically for the deposit to be cleaned out (a potential dust hazard in itself). The second disadvantage is that, if the dust is flammable, there may be a risk of a dust explosion. However, the enclosure may still be an acceptable answer if protection against explosion is incorporated in the design. Explosion protection is dealt with in Chapter 3.

5.2   PARTIAL ENCLOSURE (WITH EXTRACTION)

It may be possible to enclose the dust source but not to seal the enclosure. This may be because of lack of confidence in the short or long-term integrity of joints or covers or because apertures must be left for moving parts, controls, observation, access, etc. In such circumstances, an enclosure with the minimum total leakage area may be used and it should be kept under a small negative pressure by air extraction. The leakage will then be controlled and only inwards.

A large number of examples of enclosure designs with recommended air extraction volumes is given in ACGIH[15]. A typical example of the type of information available is given in Figure 5.1 for conveyor belt transfer point ventilation.

5.3   HOODS

Hoods are designed such that the air velocity at the point of origin of dust release is sufficient to maximise the capture of airborne dust. Where possible, hoods should be so positioned that the natural flow of material is in the direction of the hood and away from close operatives. This aspect assumes great importance where the particular material generated contains chips or particles of appreciable size and density or when dealing with a high-temperature exhaust.

Hoods may be required to move with the component or process producing the dust, such as a cutting tool. This is often achieved by the use of

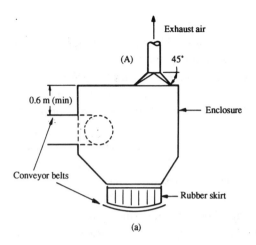

(a)

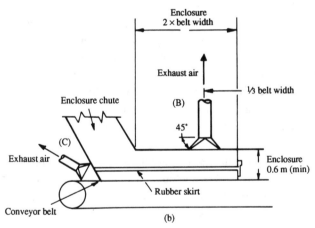

(b)

Figure 5.1   Transfer point ventilation. (a) Conveyor transfer for less than 1 m fall.
(b) Chute to belt transfer for greater than 1 m fall.

Design data:

• Transfer points should be enclosed to provide not less than 1 ms^{-1} inlet velocity at
all openings.

• Extraction point A: Minimum exhaust volume: 0.54 m^{3}s^{-1} per m width of belt for
speeds under 1 ms^{-1}; 0.75 m^{3}s^{-1} per m width of belt for speeds over 1 ms^{-1}.

• Extraction point B: Used also when fall is greater than 1 m: (same volumes as for A).

• Extraction point C: Used also for dusty materials: 0.35 m^{3}s^{-1} for belt width 0.3–0.9 m;
0.5 m^{3}s^{-1} for belt width greater than 0.9 m.

Note: Dry, very dusty materials may require exhaust volumes 1.5–2 times above values.

17

flexible connections, or very occasionally by telescopic joints. Detailed attention to design is essential in these cases, because a kinked flexible pipe can completely nullify the operation of the dust control plant.

With captor heads, the required capture velocities for different duties depend on factors such as the type and size of dust, the presence of air thermal currents and draughts and any initial motion imparted at the point of origin. Typical capture velocities vary from 0.3 m/s for fine fumes in still air up to 12 m/s for coarse contaminants released at high initial velocities into zones of rapid air movement. For any given duty, the hood designer must select an appropriate capture velocity based on either past experience or experimental observation.

Some typical forms of simple hood are shown in Figure 5.2. This also shows the air volume required to achieve the capture velocity at a given distance from the hood face.

In assessing the potential effect of a proposed hood design, it should be appreciated that the velocity of the air moving towards the hood reduces as a function of the square of the distance from the hood. In practical terms, this means that the capture velocity of the hood face has to be very high if the effect of the hood is to be felt more than, say, 0.5 m away. A corollary of this is that even minor disturbances of the air pattern, caused by draughts, for example, can severely disturb the effectiveness of hood capture if the design does not allow for this.

In some applications, it is desired to minimise air flow from behind the hood face as this may be a zone where no contamination exists. This can be achieved by providing a wide flange around the hood face. This also has the effect of improving the effectiveness of a given hood in its forward regions and helps to reduce energy consumption.

For a *given extraction volume*, the capture velocity at any given point in front of a hood becomes progressively less dependent on hood size as the point moves further away from the hood face. Large hoods, however, have two advantages; their zone of influence is greatly increased and they possess a lower entrance loss and hence reduced power requirement. The chief technical disadvantage of large hoods is that their capture velocity is low. This can be detrimental when dealing with coarser dusts which are likely to settle rapidly unless they are sucked quickly into the hood.

In large hoods a slotted face may be necessary to achieve uniform distribution of induced air, particularly if the duct take-off point is to one side.

Flow rates can be determined
using figure 5.2

Hood type	Description	Aspect ratio, $\frac{W}{L}$	Air volume
	Slot	0.2 or less	$Q = 3.7\,LVX$
	Flanged slot	0.2 or less	$Q = 2.8\,LVX$
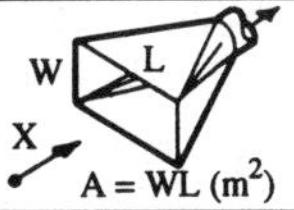 $A = WL\ (m^2)$	Plain opening	0.2 or greater and round	$Q = V(10X^2 + A)$
	Flanged opening	0.2 or greater and round	$Q = 0.75V(10X^2 + A)$
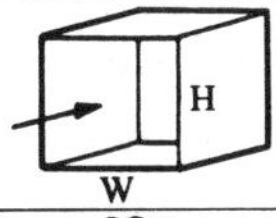	Booth	To suit work	$Q = VA = VWH$
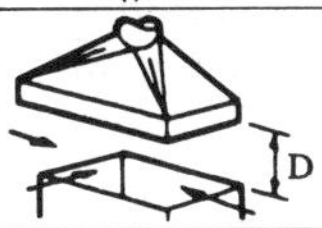	Canopy	To suit work	$Q = 1.4PDV$ P = perimeter of work D = height above work

Figure 5.2   Common types of simple hood and their recommended extraction characteristics. Q = hood extraction volume (m³/s); V = centre line capture velocity (m/s); A = area of hood face (m²); X = distance from hood face (m); L = length of hood slot (m); W = width of hood slot (m); and H = height of booth (m).

Slot velocity does not contribute towards capture velocity; the purpose of the slot is purely to provide sufficient pressure drop to prevent all the air being induced from that section of the hood nearest the suction off-take.

It cannot be over-emphasised that good hood design requires the most careful observation and meticulous attention to detail. For many common industrial operations producing dust or swarf proven hood designs exist together with recommendations for their associated entrainment rates. Advice on specific applications can be obtained from organisations such as the British Cast Iron Research Association, British Steel research centres or the Organic Research Association. Further advice is given in References 13–21. Hoods are also often supplied with machines by the machine manufacturers. However, it must be recognised that standards tend to improve with time and the influence and authority of the various statutory inspectorates must be taken into account.

For the design of receptor hoods, on-site measurements of the size and shape of the dust cloud (using, for example, a light beam) and the velocity of approach of the contaminated air (over the complete cross section of the dust cloud) are usually required, together with other measurements such as gas temperature, etc. Based on these observations, the size, shape and position of the hood can be decided and the required volume of exhaust air calculated. It is not sufficient merely to have to correct air volume; careful design of the hood is essential, with particular attention required to the distribution of air both across the hood face and within the hood itself. Clearly, the exhaust system must extract air from each position on the hood face at least as fast as contaminated air approaches it, otherwise there will be a spillage or escape of fume from the hood.

For more complex arrangements and specialised process applications it is often necessary to consider conducting a smaller-scale, model test. This is meaningful, because within a wide range the shape and distribution of velocity contours and streamlines are the same, regardless of the size of hood or the amount of gas entering the opening. Provided the hood shape is geometrically similar, a model hood will exhibit the same contours, expressed in terms of the hood opening face velocity, as the full-scale hood. Two qualifications are advisable; it is unwise to use a linear scale model smaller than 1/15 full scale and it is difficult to reproduce accurately on a model such phenomena as high-temperature emissions and spray patterns.

6. DUCTING

Ducting is required on any dust control system where the collector and fan are located at some distance from the points of dust generation and control. In those cases where the collector can be mounted directly on, or inserted into the machine generating or receiving dust, the use of ducting can be avoided, a welcome economy from many aspects. Unit collectors mounted adjacent to the dust source can also avoid the need for long lengths of ducting.

6.1   DUCTING DESIGN PRINCIPLES

Ducting should be sized to give a constant conveying velocity throughout the transport system. For particulate systems, the gas velocity in the ducting should never be allowed to fall low enough for significant dust settlement to occur. On the other hand, excessively high velocities are wasteful of power and may cause accelerated abrasion and more noise. As ducts seldom operate at a single gas through-put condition, the designer must consider both extremes of velocity to ensure that he selects an optimum design velocity that will be acceptable for all known plant conditions.

The duct design velocities that are used in practice are generally higher than would be predicted from consideration of Stokes' settling velocities for particles of a given size and density.

For the 'average' industrial dust, a 'rule of thumb' conveying velocity employed in preliminary designs is usually around 18 m/s. Fumes will convey at velocities as low as 5 m/s, whereas heavy or moist dusts will require 25 m/s. When finalising the duct design, it is necessary to seek an optimum duct diameter taking into account both capital and operating costs. Smaller ducts are not only cheaper to install but they minimise the effect of reduced suction conditions, local condensation and dust agglomeration. With heavy, expensive ducting it is sometimes preferred to accept intermittent periods of high pressure loss and to design the duct to produce velocities as high as 40 m/s. It is always worthwhile to prevent unnecessary friction losses by keeping the duct interior smooth and employing bends of generous radii.

6.2   MECHANICAL FEATURES OF DUCTS

Most plants have the fan located at a point just prior to the cleaned air discharge. This ensures that should any leaks develop, ambient air will be drawn into the system rather than contaminated air blown out. Ducting should be made as airtight as possible, and this will require careful selection of flange gasket materials and may also involve external sealing of the joints. On the other hand, with smaller ducts slipjoints are often adequate and greatly facilitate neat erection.

Connections to the main duct should preferably be made from above or, if that is not possible, at the side. This avoids any tendency for material to settle in a branch line when this leg is turned off or is subject to a low conveying velocity.

Branch ducts should enter the main at an angle of 30° (preferred) to 45° and two branches should not enter directly opposite each other. Where the branch enters the main, there should be a gradual increase in the diameter of the main to take the increase in gas volume. Guidance on good ducting design practice is given in ACGIH[15].

Even with a well-designed system there are likely to be occasions when a some fall-out of dust will occur. When deciding on the ducting layout it is sometimes possible to include vertical or sloping sections which can act as convenient drop-out areas. However, controlled deposition is difficult to achieve reliably and the ambitions of the designer may not be fully realised in practice. Inspection doors, hinged and clipped with effective sealing, should be placed at intervals along the ducting, say at 3 m pitch, adjacent to bends where dust build-up or deposition may occur. Most ducting is overhead and must be adequately supported. The duct supports and the ducting must be capable of bearing the additional weight incurred if an extensive dust fall-out occurs within the system.

For the handling of air carrying 'nuisance' dusts, the ducting should typically have a wall thickness of 3 mm; where the dust loading is high or where the dust is abrasive, the wall thickness on bends might be increased to 5 mm. For ventilation systems or systems carrying very light dust loads wall thicknesses of less than 3 mm might be used. Normally, ducts carrying hot or corrosive dust-laden gases from dryers, furnaces, etc, will have a minimum wall thickness of 5 mm. Large ducts are usually made of carbon or stainless steels with a minimum wall thickness of 6 mm.

6.3   MULTIPLE DUCT SYSTEMS

Individual consideration of the major points of dust generation and dispersion leads to an estimate of the air entrainment flow-rate required at each point, the summation of which results in the total air entrainment capacity for the complete plant. As most collection systems operate below atmospheric pressure, it is often the practice to add 10% on to this assessed air rate to accommodate any loss of effective suction through leakages in either the ducting system or the collector.

A reliable assessment of this air volume capacity is of fundamental importance to the effectiveness of any nuisance-dust control plant. Admittedly, failure to achieve total control is not unusual, and in any case such perfection could be prohibitive in cost, but any serious shortfall in air entrainment capacity can only be rectified at high cost and inconvenience. Limited or local shortfall in capacity can often be countered by improving the design and location of the existing hoods and by adjusting their relative extraction rates. Such changes are often obvious and desirable when the plant is first observed in operation.

When extracting from several hoods and/or enclosures into a common duct, it is necessary to 'balance the system' so that each of the various branch ducts delivers the intended air volume at the required conveying velocity. Without balancing, the total air volume would distribute itself automatically according to the resistance of the available flow paths with the bulk of the flow passing through the line of least resistance. Two methods are available for balancing the system, the first being by far the more widely used and the more flexible.

METHOD 1 (AIR BALANCING USING DAMPERS)
This method depends on the use of dampers which are adjusted during commissioning to achieve the desired air volume in each branch duct. The dampers should be of a type and in a location that will discourage subsequent unauthorised tampering. The design calculation begins at the branch of greatest resistance using the minimum conveying velocity. Pressure drops are calculated through the branch and through the various sections of the main duct up to the fan. At each section of the main where a branch joins, the desired volume of flow from the branch duct is added to the volume in the main and the main duct diameter increased accordingly. The same guidelines apply where two or more branch ducts join prior to linking in with the main. Care must be exercised in choosing the branch of greatest resistance initially; if the choice is incorrect, any branch having a higher resistance than that chosen will fail to deliver the desired volume even with its damper fully open.

METHOD 2 (AIR BALANCING WITHOUT DAMPERS)

This method is normally used (and is often mandatory) where dangerous dusts are being handled, in order to safeguard the system against tampering with dampers and the possibility of dust accumulation in the branch ducting caused by damper obstruction. The design calculation also begins with the branch of greatest resistance and proceeds branch to main and section of main to section of main up to the fan. Using this method, however, at each junction of branch and main (or branch and branch), the static pressure necessary to achieve the desired flow in the branch must be made equal to the static pressure in the main. The static pressures are matched at the desired rates of flow by suitable choice of duct diameters, elbow radii, etc.

Worked examples, showing both methods of calculation, are given in References 15 and 20. One of the most common errors in designing multiple-duct systems is to incorporate too many extraction points into the one system. Normally, a total of 10 points should be regarded as a maximum.

7. PLANT SELECTION PROCEDURE

Many factors must be considered when approaching the critical stage of selecting a particular type of dust and fume collector. The ultimate choice will be governed by a blend of technical, commercial and legal criteria plus a liberal quantity of sound common sense.

Before selecting a collector system the user must have fully considered the matters raised in the previous chapters of this guide. To choose a dust collector system without a full knowledge of both the dust and gas involved can lead to severe technical shortcomings and will almost certainly mean a greater expense than would otherwise be necessary.

The selection procedure for an appropriate collector will not normally be achieved in a single step. Having made his initial choice, the potential user must then consider the ancillary items, such as fans, dust disposal plant, water treatment, etc. Sometimes a difficult, even insurmountable, problem may arise which will necessitate revising the choice of collector. This optimisation process continues until all reasonable requirements can be met both technically and economically.

If there is previous experience in applying a certain type of collector to a given application then this will obviously be a valuable factor when a repeat plant is being considered. Care needs to be exercised in comparing applications, however, due to the fact that even small changes in plant design or operation can lead to significant differences in the performance of the dust collector.

It is worthwhile for the potential user to cast his net as widely as possible to ensure that he finishes with the optimum selection for his particular problem. As technology advances designers find ways of overcoming previous limitations and the choice of dust collectors widens.

7.1 INFORMATION REQUIRED

The following paragraphs summarise the type of information normally required to assess the type and design of equipment best suited to solve the problem. Not all the points are essential for all types of equipment but, in order to produce a comprehensive specification, as many should be used as is practical.

CONSIDERATION OF THE PROCESS

- Description of the process and any relevant or unusual features.
- Details and size of the actual dust-producing equipment including drawings, if available.
- Mode of operation of plant — continuous or batch (giving cycle times).
- Description and analysis of raw material, or fuel, and any expected variations where these may affect dust produced.
- Description, rating and performance of any existing dust collector and ancillary equipment.

OPERATING CONDITIONS

- Gas volume, pressure and temperature, including likely variations (normal, maximum, minimum, start-up conditions, etc).
- Gas analysis and density of gas.
- Moisture content or dew point of gas.
- Barometric pressure or elevation of plant site.
- Maximum allowable pressure drop through new equipment.
- Other operating conditions or constraints on design (such as available space, power supply, fan capacity, environmental considerations).

DUST CHARACTERISTICS

- Nature of dust to be collected, including chemical analysis if available.
- True density and bulk density of dust.
- Unusual characteristics of dust (hygroscopic, pyrophoric, sticky, abrasive, electrostatic, etc).
- Toxicity of dust.
- Particle size analysis and method used for obtaining it; particle shape.
- Inlet dust concentration, normal and peak.
- Wetability.
- Dustability.
- Resistivity of the dust, if available.
- A sample of dust should be provided, but care must be taken in the sampling and storage of this material to ensure that it is representative of *in situ* conditions.

PERFORMANCE REQUIRED

- Minimum dust collection efficiency, and/or
- Maximum allowable outlet dust concentration or presumptive limits if applicable.

SITE LAYOUT

- A sketch or drawing should be provided showing the dust source and the space available where the dust and fume equipment could be located. This sketch/drawing should also show the headroom available and any possible obstructions. The availability of services such as power, water and drainage should be incorporated. It may also be necessary to show the geographical location of the site with respect to nearby buildings, residential and topographical features.

COLLECTION OF REQUIRED DATA

There are five principal methods of obtaining design data:

- For new dust collectors on existing processes, measurement of temperatures, flowrates and dust sampling in a duct or chimney can provide accurate and direct data.
- For dust collectors on new plants, not yet built, careful comparison with existing processes can provide valuable data, providing that appropriate scaling is made and allowances made for differences in operation and raw materials.
- For completely new applications, where no previous experience is available, it is often possible to calculate gas flowrate and temperature from the quantity and composition of reactant materials — this is particularly true for combustion and chemical process applications.
- Pilot plant testing can provide valuable design data on new applications where existing design data is limited.
- In ventilation applications, the definition of flowrate is inextricably linked with overall design of the hooding and ventilation system and it may be necessary to consult reputable equipment suppliers in assessing the design parameters of the gas cleaning plant.

Chapter 18 gives general advice on recommended experimental techniques including a brief survey of the sampling and test methods used to obtain design data.

7.2   FURTHER CONSIDERATIONS

PROCESS COMPATIBILITY

Considering the gas cleaning plant in isolation from the main production process and the general environment can be at best unwise or at the worst catastrophic. Conversely, by carefully studying, for example, the cyclic variations in the emission of dusts from a particular process, the potential user may be able to reduce the size and cost of the dust control scheme significantly, because, although the collector must be capable of accepting the instantaneous mass of dust entering the plant, it may be acceptable to design the regeneration, storage and dust disposal systems on the average rather than the peak dust loading.

Another factor to be considered is the marriage of skills required to operate and maintain the main process and the dust control plant. Choosing sophisticated equipment for a site which employs totally unskilled personnel clearly can lead to both plant and labour difficulties.

LOCATION

The location of the process plant site will already have been taken into account when considering the maximum allowable emission permitted from the plant. It is not always necessary or even desirable, however, for the dust control plant to be sited next to the main production process. In extreme cases, they can be as much as several hundred metres apart. Of course, remote location of the dust collector introduces a number of additional factors which must be considered. Is dust or condensation likely to be deposited in the transfer duct? What is the temperature fall along the duct? Can plant instrumentation and control be effected reliably and safely? Can the plant operator see the result of his attempts to produce a clean exhaust? Can the emptying of dust or slurry containers be effectively controlled? Even if these additional factors introduce difficulties of their own, there are often real advantages in locating the dust collector remotely. The absorption of high temperature peaks along a long transfer duct and the isolation of employees from fan noise are just two possible benefits.

UNIT DESIGN

Many designs of dust and fume collectors are available as packaged assemblies. Such units can offer the advantage of low purchase cost, minimum site installation requirements and short delivery time. The buyer should ensure, however, that the standard unit is really suited to the required duty. In no circumstances should the purchaser attempt arbitrarily to select a dust collector from a sales brochure without first defining a detailed specification of the technical duty. Fabric filters are the most frequently used collectors of unit design and these are discussed in more detail in Chapter 11.

SUCTION OR PRESSURE OPERATION

Most dust and fume collectors operate close to atmospheric pressure. They are usually fitted with an induced-draught fan which is designed to overcome the resistance of the hoods, ducts and gas cleaning plant. Such a system has two inherent advantages. First, most of the equipment operates under a slight suction so that any leaks are inwards. This is particularly beneficial with gases or dust that are toxic or otherwise harmful. Secondly, the location of the fan on the clean side of the collector prolongs the life of the fan impeller and requires less maintenance. This also enables higher-efficiency fans to be used with a consequent saving of energy.

Sometimes, however, it is preferable, or necessary, to operate the dust collector under a positive pressure. In some instances, the clean gas discharge from the gas cleaning plant might be at a sufficiently high level to obviate the need for a separate stack. Alternatively, pressure operation may mean simpler mechanical construction for the casing of the dust collector, or fewer problems with air ingress through leaking rotary valves.

When choosing between pressure and suction operation a practical point to remember is that it is both difficult and expensive to make large, flat-sided plants which are entirely leakproof. For leaking inwards, with suction operation, the fan should be sized to accommodate the ingress of air. With pressure operation, outward leakage means that it is necessary to ensure that neither dust nor the process gas is hazardous to plant operators.

COLLECTORS IN SERIES

Some gas cleaning plant can be considered to be inherently multi-staged; for example, plate-type wet washers and electrostatic precipitators. The total efficiency of such collectors is represented by the cumulative effect of the individual stages, thus the more stages the better the collection efficiency.

On the other hand, there is little advantage to be gained, in terms of collection efficiency, in using several bag filters in series, although cyclones may be arranged in series if dealing with materials which easily form agglomerates. However, collectors in series can also be employed for other reasons. Cyclones or inertial separators might be installed upstream of other collectors, such as bag filters, to remove the coarse particles or perhaps to protect the fabric

29

from sparks. A low-energy scrubber is often installed upstream of a higher energy scrubber with the main purpose of cooling and saturating the incoming gas. Demisters are required downstream of a wet washer to minimise carryover of entrained wash-liquors. In situations where a high reliability is required collectors might be installed in series, so that if the first collector fails the second is available to perform the necessary separation. These are just a few examples which illustrate that many advantages can be gained by careful interplay of collectors, sometimes to improve overall collection efficiency but also to give increased reliability. The user should beware, however, that sometimes series operation of dust collectors can be disadvantageous. For example, a mechanical pre-collector, which separates out only the coarse particles, may present the main collector with a gas stream containing only the remaining fine particles. Without the assistance of the larger particles the fine material might possess characteristics which make collection or handling more difficult than when the dust has a wide spectrum of particle sizes.

7.2.6   METHODS OF GAS COOLING

In the field of dust and fume control, gas cooling is a subject which is receiving much wider attention than in the past because of the increasing importance of energy recovery and the need for better control of gas temperatures, particularly on the more difficult processes. Gas cooling also allows a wider choice of dust collector to be considered. It should, however, be remembered that with all forms of gas cooling there is a risk in cooling gases below or near their dewpoint, particularly if acidic constituents are present which, when combined with free moisture, can cause severe corrosion. The different methods of cooling available can be subdivided as follows:

Cooling with heat recovery: Waste-heat boilers;
                               Recuperators and regenerators;
                               Forced-draught coolers.

Indirect cooling: Water-cooled ducts;
                     Trombone or U-tube coolers;
                     Forced-draught coolers.

Direct cooling: Evaporative cooling;
                   Quench cooling;
                   Dilution air cooling.

It should be noted that heat recovery systems, whilst appearing attractive in theory, do not always prove to be compatible with dust collection systems. Capital and maintenance costs may be high and a process shutdown may mean an inconvenient loss of heat supply to users. The return on the capital cost of waste-heat boilers, recuperators and regenerators will generally be attractive only in situations where large volumes of gas are available at high temperatures.

Water-cooled ducts are normally used only with high temperatures as they are not very effective below 700°C. Trombone or U-tube coolers have been widely used on gas cleaning applications. This type of cooler consists of a series of large diameter carbon steel tubes, mounted vertically, heat transfer taking place by radiation and natural convection to the ambient air. Consequently, they are not very effective below 250°C because of low heat transfer coefficients. The major disadvantage of this type of cooler is the lack of control over the degree of cooling obtained. They are, however, relatively easy to clean of dust deposits.

In the forced-draught cooler, the usual arrangement is that ambient cooling air is forced through horizontal tubes by a number of axial flow fans whilst the process gases pass vertically over the outside of the tubes. Good control over the outlet gas temperature can be achieved by switching on and off one or more cooling fans. Gases can be cooled from 500°C to as low as 135°C although the use of alloy steel tubes permits hotter gases to be handled. The cooling air can be used for space heating.

If evaporative cooling is used prior to a dry collector, special care is needed in the design and operation of the spray control system to ensure good temperature control and to avoid carryover of water droplets, especially during the starting up and shutting down of the plant. The main disadvantage of this method is the increase in the dewpoint of the gas which can cause condensation problems and blinding of the filter media where fabric or deep bed filters are being used. Quench cooling differs from evaporative cooling only in that there is no need to control accurately the quantity of spray water. This technique is used only with wet scrubbing.

The above methods lead to a reduction in the volume of gas to be treated by the dust collector, and savings on the size of the collector and ancilliary equipment may offset the cost of the cooling system. Instead of a cooler, dilution air alone can be used; however, this is often not cost-effective because of the very large increase in the volume of gas to be cleaned. Use of dilution air, therefore, is normally restricted to a modifying function such as for partially

31

cooling hot process gases before entry to a cooler to allow carbon steel to be used or for final control of gas temperature prior to entering the filter. Dilution air can also be combined with evaporative cooling to reduce the likelihood of dewpoint problems. Further details on the cooling of gases prior to fabric filtration are given in Reference 28.

7.3   A GUIDE TO PLANT SELECTION

The proposed selection procedure consists of two stages. The initial selection factors are shown in Table 7.1 on pages 34 and 35. When the initial choice has been made, this may indicate one or two types of collector suitable for the proposed duty. Table 7.2 on pages 36 and 37 should then be consulted for a more detailed technical evaluation of the likely process conditions. If all the conditions cannot be successfully accommodated within the plant initially selected from Table 7.1, then it may be necessary to reconsider the primary factors. For example, the inlet gas could be cooled to a different temperature upstream of the collector. Alternatively, user preferences may have to be reconsidered in order to arrive at a technically acceptable solution.

Only the more common types of collector have been covered in this chapter. New forms of equipment are continuously being developed, but major technology advances which have a significant effect on the overall usage pattern are rare. It is suggested that inexperienced users should restrict their selection to the well-tried devices listed, unless there is adequate information available on the performance of alternative equipment specific to the user's application.

7.3.1   PRIMARY FACTORS

COLLECTION EFFICIENCY VERSUS PARTICLE SIZE

The most common method of characterising the performance of a dust collector is by means of a 'grade-efficiency' curve which is a plot of collection efficiency versus particle size. A typical example is shown in Figure 7.1. 'Typical' grade-efficiency curves, based on full-scale plant performance tests, are available in the literature for a wide range of collectors. Using these curves, it is possible to predict the overall collection efficiency for a particular dust collector on a given application, provided that the size analysis of the inlet dust is known. The method of calculation is given in Reference 26. Extreme care should be exercised, however, in the use of these curves and the user's attention is drawn to the following points:

• The size analysis of the inlet dust should be based on the same particle 'diameter' as the grade-efficiency curve (eg Stokes' diameter) — see Chapter 18.

• The grade efficiencies are likely to be affected by particle density, particle shape and by the degree of particle agglomeration. Frequently, the size distribution *in situ* is very different from the 'dispersed' size analysis obtained by one of the standard particle sizing techniques.

• The grade-efficiencies, for scrubbers particularly, will depend on the operating pressure drop. With high-energy scrubbers, a series of grade-efficiency curves is required.

• The grade efficiencies will be affected by the physical condition of the air or gas stream, eg temperature, humidity, etc.

Despite the above difficulties, grade-efficiency data will allow the user to make an initial choice of the type of collector that might be used.

Based upon the published grade-efficiency curves, Table 7.3 on page 38 gives approximate collection efficiencies for a range of dust collectors for arbitrary particle diameters of 10 μm, 5 μm, 2 μm and 1 μm. Clearly, inertial collectors, cyclones and low-pressure-drop scrubbers should not be used where the efficient collection of fine dust is required (unless the inertial collectors or cyclones are being used as precleaners). With high-energy scrubbers, fine

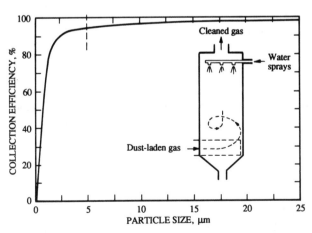

Figure 7.1   Typical grade efficiency curve for a spray tower (efficiency at 5μm is 94%).

TABLE 7.1
Primary factors for the selection of plant for dust and fume control

	Smallest size particle to be collected		
	> 10 micron	1–10 micron	Sub micron
Cyclones	✓	Care	Beware
Wet washers, low energy	✓	Care	Beware
Wet washers, high energy	✓	✓	Care
Dry electrostatic precipitator	✓	✓	✓
Wet electrostatic precipitator	✓	✓	✓
Aggregate filters	✓	✓	Care
Fabric filters	✓	✓	✓
Fibrous filters	✓	✓	Care

Key
✓   Can generally cope with the process requirements if well designed.
Care   Special attention required in plant design and operation to prevent problems.

particles can be collected with high efficiency, but the energy requirement increases exponentially in the submicron range as the particles become finer. The positive nature of fabric filters will generally ensure a high collection efficiency over a wide particle size range provided that the filter cloth has been correctly selected and conditioned. Similarly, fibrous and aggregate filters will capture the finest dusts if fitted with beds of appropriate quality. With electrostatic precipitators, collection efficiency is insensitive to particle size. The desired efficiency can be obtained by increasing the surface area of the collector electrodes and/or the number of electrical 'fields'. There are many industrial applications of both dry and wet precipitators collecting submicron dusts.

TABLE 7.1 (continued)
Primary factors for the selection of plant for dust and fume control

Gas temperature inlet to collector				User preferences (if practical)			
> 400°C	250°C–400°C	> dewpoint up to 250°C	Near dewpoint	Dry product	Low initial cost	Low operating cost	Minimum technical complexity
✓	✓	✓	Care	✓	✓	✓	✓
Care	✓	✓	✓	Beware	✓	✓	✓
Care	✓	✓	✓	Beware	Care	Unlikely	Care
Care	✓	✓	Care	✓	Unlikely	✓	Care
Beware	Care	Care	✓	Beware	Unlikely	✓	Care
✓	✓	✓	Care	✓	Care	Care	Care
Beware	Care	✓	Care	✓	Care	Care	Care
Beware	Care	✓	Care	Beware	✓	✓	✓

Beware   Could lead to severe operational difficulties; alternatives avoiding the problem are normally sought.

Unlikely   On purely economic grounds, alternatives generally favoured if suitable.

GAS TEMPERATURE

The temperature of the gas to be cleaned can affect the choice of plant especially as far as the materials of construction are concerned. Cyclones, for example, can be made from most heat-resistant materials and are thus widely used for high-temperature applications. With fabric filters, however, gas temperatures are in general limited by the heat sensitivity of the media. The maximum temperature that each fabric material can withstand cannot be precisely stated because it is also a function of exposure time and other operating conditions. As a guideline, however, natural fibres are usually limited to applications below 80°C; for temperatures up to 200°C, appropriate synthetic fibres are used, whilst glass fibre

35

TABLE 7.2
Secondary factors for the selection of plant for dust and fume control

	Dust properties						
	High inlet burden	Erosive	Sticky	Light fluffy	Difficult to wet	Pyrophoric	Resistivity problem
Cyclones	✓	✓	Beware	Beware	✓	Care	✓
Wet washers low energy	✓	✓	✓	✓	Care	✓	✓
Wet washers high energy	✓	Care	✓	✓	Care	✓	✓
Dry electrostatic precipitators	Care	✓	Care	Care	✓	Care	Beware
Wet electrostatic precipitators	Care	✓	✓	✓	Care	✓	Care
Aggregate filters	Care	✓	Beware	✓	✓	✓	✓
Fabric filters	✓	Care	Beware	✓	✓	Beware	✓
Fibrous filters	Beware	✓	Care	✓	✓	Care	✓

Key
✓   Can generally cope with the process requirements if well designed.
Care   Special attention required in plant design and operation to prevent problems.

TABLE 7.2 (continued)
Secondary factors for the selection of plant for dust and fume control

Gas conditions				Other factors		
Constant pressure drop	Varying flow	Explosive, combustible *	Corrosive	Suitable for high pressure	Minimum ancillary equipment	On-line regeneration
✓	Care	Care	Care	✓	✓	✓
✓	Care	✓	Care	✓	Care	✓
✓	Care	✓	Care	✓	Care	✓
✓	Care	Beware	Care	Care	Care	✓
✓	Care	Beware	Care	Care	Care	Care
Care	✓	Care	Care	✓	Care	Care
Care	✓	Care	Care	Care	Care	Care
Care	Care	Care	Care	✓	✓	Beware

Beware	Could lead to severe operational difficulties; alternatives avoiding the problem are normally sought
*	For all gas cleaning problems associated with explosive or combustible materials, competent advice should be sought.

37

TABLE 7.3

Approximate collection efficiencies* (%) of dust collectors on different particle sizes

Type of collector	Efficiency			
	at 10 μm	at 5 μm	at 2 μm	at 1μm
Inertial collector	30	16	7	3
Medium-efficiency cyclone	45	27	14	8
High-efficiency cyclone	87	73	46	27
Low resistance cellular cyclone	62	42	21	13
Tubular cyclone	98	89	77	40
Irrigated cyclone	97	87	60	42
Self-induced spray deduster	98	93	75	40
Spray tower	97	94	87	55
Wet impingement scrubber	> 99	97	92	80
Disintegrator	99	98	95	91
Venturi scrubber — medium energy	> 99.9	99.6	99	97
Venturi scrubber — high energy	> 99.9	99.9	99.5	98.5
Electrostatic precipitator	> 99.5	> 99.5	> 99.5	> 99.5
Irrigated electrostatic precipitator	> 99.5	> 99.5	> 99.5	> 99.5
Shaker-type fabric filter	> 99.9	99.8	99.6	99
Pulse-jet fabric filter	> 99.9	> 99.9	99.6	99.6

* For dust of density 2700 kg/m^3

will withstand continuous operation at 260°C. There is also limited experience with both mineral and metal fibre media which are claimed to withstand temperatures up to 600°C. However, inexperienced users would be advised to restrict their initial choice to conventional fabrics where a greater depth of plant information is available. Higher temperatures are common with fibrous filters, where metallic fibre pads will operate at 500°C, and with gravel-bed filters.

Wet dedusters are used at gas temperatures up to 1200°C but at elevated temperatures they have a higher water consumption because of evaporation within the unit. Hot gases are cooled by wet dedusters and this results in a less buoyant and more visible plume. Sometimes, reheating is employed to avoid stack condensation and to give improved plume buoyancy[37].

High-temperature operation of electrostatic precipitators is possible, but in practice they are rarely designed for operation above 450°C as special steels and insulator materials are required, making it economically unattractive. Precipitators sometimes work better within certain temperature ranges due to the temperature effect on dust resistivity.

An alternative to the choice of special materials of construction or expensive fabrics to withstand high process gas temperatures is to cool the gas. Methods of gas cooling are discussed in Section 7.2.

WET VERSUS DRY COLLECTION
An early decision that has to be made when considering a gas cleaning problem is whether to collect the particulate material in a dry state or as a slurry. Often the main process may dictate the choice, but in other circumstances the selection is made by the gas cleaning plant designer. In general, the initial preference should be for a dry collector, since this avoids both the supply and operation of a water treatment system. However, in the presence of high temperatures, acidic components or sticky dusts, some form of wet collector will often be the correct choice. It may also be easier to incorporate low-temperature heat recovery systems when employing a wet washer because the direct contact between hot gas and the scrubbing liquor effectively recovers all useful heat from the gas. A further comparison of the relative advantages and disadvantages of wet collectors is made in Chapter 9.

LOW-COST SOLUTIONS
The economic evaluation of competitive gas cleaning plant is dealt with more fully in Chapter 17. For the present purpose, cyclones, fibrous and fabric filters and certain types of wet washers may be considered as low-capital-cost plants because they are widely available in the form of a standard package which may conveniently meet the user's requirements. It must be stressed, however, that the total cost of the installed plant must be considered. This should include all ancillary equipment as well as civil, electrical and erection work.

In Table 7.1 cyclones, fibrous filters and most electrostatic precipitators have also been considered to be low-operating-cost devices, because they are often associated with a low operating gas pressure drop. In the more detailed evaluation, it would be necessary to consider other operating cost factors including labour costs, pumping costs, spares, maintenance requirements, etc.

TECHNICAL COMPLEXITY

No gas cleaning plant can really be considered as a 'fit and forget' device. However, cyclones and fibrous filters do not usually require high-grade supervision. Other types of plant often require a degree of 'user experience' to achieve their optimum performance over a long period of operation. The weighting in this column of Table 7.1 takes account of the amount of maintenance that the plant may require during its operating life.

7.3.2   SECONDARY FACTORS

HIGH INLET DUST BURDENS

Cyclones and some types of wet washers are capable of operating with inlet dust burdens in excess of 200 g/m^3. However, the maximum inlet concentration to these devices is clearly linked to the efficiency which they can achieve on the dust (a function of its size) and the required emission concentration.

A fabric filter is normally limited, by pressure drop considerations, to maximum inlet dust concentrations of around 50 g/m^3 but higher loadings may be acceptable with continuously cleaned filters if the bulk of the incoming dust is directed away from the filter tubes.

If the inlet burden to an electrostatic precipitator exceeds 20 g/m^3, difficulty may be experienced with electrical stability. This effect is usually counteracted by selecting a discharge electrode of suitable characteristics.

Fibrous filters are normally used as dust collectors only when the incoming burden is considerably less than 1 g/m^3. Pre-collector cyclones can be introduced upstream of filters and precipitators to reduce the inlet burdens to these devices.

DUST DENSITY

With inertial separation devices, such as cyclones and many wet scrubbers, the efficiency of collection increases for higher density particles. Other types of collector are less dependent on particle density, although care should always be exercised to ensure that heavy agglomerates do not prematurely deposit at inconvenient points in the inlet casing.

Light fluffy dusts require special consideration with most types of collector. Cyclones are usually inefficient with such materials, and fabric filters and electrostatic precipitators should be designed for low velocities to avoid re-entrainment of collected material. When handling light dusts with wet collectors, there is the added complication of a low settling velocity in the liquor treatment sedimentation systems.

STICKY DUSTS
Materials which are inherently glutinous or hygroscopic will blind most positive types of filter. Even devices with large gas openings such as cyclones and dry precipitators can suffer deposition problems if the collected dust is not free-flowing. Generally, these conditions favour wet collectors although, even then, it is important to avoid material build-up at wet/dry interfaces.

EROSIVE DUSTS
Severe wear problems may occur in dust collectors with local high gas velocity characteristics when erosive dusts are present. In these cases, it is preferable to avoid cyclones and high-energy venturi scrubbers unless specially wear-resistant linings can be accommodated. With bag filters, such conditions favour the arrangement whereby the dust is collected on the outside of the bag as the problem of wear on the bag entry mouth is then avoided. Electrostatic precipitators are well suited to erosive applications as they are low-velocity devices, but any internal moving pans are likely to have reduced life under these conditions.

DIFFICULT-TO-WET-DUSTS
Certain dusts, particularly those of an organic nature, may be difficult to wet when in a finely dispersed form. In severe cases, this may cause lower than anticipated collection efficiencies in wet collectors.

COMBUSTIBLE DUSTS
Dusts which are easily ignited or gas streams which carry sparks or even flames into the dust collector are obviously a hazard when flammable materials are present. Fabric filters are particularly prone to severe damage in these conditions. All dry collectors need special care to avoid hopper fires and in these it is advisable to minimise the accumulation of dust anywhere within the collector. Special provision may be necessary to isolate the collector from the rest of the

41

extraction system. Wet collectors, if otherwise acceptable, are intrinsically safer for highly combustible materials.

In all cases a full knowledge of the risks and prevention of fires and explosions in dust handling systems is vital to safe operation (see Chapter 3).

DUST ELECTRICAL PROPERTIES

The electrical properties of a dust are most important to an electrostatic precipitator but can also influence the selection of other dry collectors. For efficient precipitation, the dust resistivity should lie in the range 10^5 to 10^{11} ohm cm. By adjusting the gas temperature, or moisture content, most dusts fall within these limits (see Chapter 10).

Highly resistive dusts can accumulate high electrostatic charges and this can give rise to undesirable effects in fabric filters as well as in electrostatic precipitators. Fabric filter cloth can be obtained which has a conductive element woven into the material and, in conjunction with an effective earthing system, this will reduce the risk of spark discharge.

TOXIC DUSTS

When handling toxic dusts, the collector should preferably operate under suction conditions so that any leakage is inwards rather than outwards towards the operators. Special precautions are required for safe working and the design should permit as much maintenance from the clean side of the collector as possible. Mechanically simple equipment such as cyclones and fibrous filters provide an obvious advantage in this respect.

Most collectors will be associated with a dust or slurry handling system. Equal care is needed in the design of this section of the plant to ensure that the material is transferred into sealed containers without leakage.

RESISTANCE TO GAS FLOW

As most dust collectors have a fixed geometry their pressure drops will be proportional to their operating gas velocity, usually with a square law relationship. Some types of wet washer, however, can be designed to present a variable area to gas flow to maintain constant pressure drop over a wide flow range. The resistance to gas flow may also vary with the amount of dust deposited within the collector.

Dust and fume collectors which have a true 'filtering' action, wherein the gas passes through a permeable membrane which acts as a barrier to the dust, cannot normally operate at a constant pressure drop. As dust is deposited in fabric, fibrous and aggregate bed filters the resistance to gas flow will increase. This leads to an increasing pressure drop up to a limit determined by the process designer. At that point it is necessary either to replace or to regenerate the filtering surface, thus giving a step reduction in the gas pressure drop.

This cyclic variation in pressure reacts on the fan and in some cases will lead to variations in the gas volume displaced by the fan. Thus, the volume of gas extracted from the process plant or hood may alter with time, and to counteract this additional damper or fluid-coupling control of the fan may be required.

This degree of cyclic variation does not occur with pulse-jet or some multi-unit fabric filters which have a large number of sections at varying stages of fabric regeneration. Such installations can maintain a constant average pressure drop because the regeneration of one individual section is not significant to the filter as a whole.

Most cyclones, wet washers and precipitators maintain a relatively constant resistance to gas flow, irrespective of changes in the quantity of dust entering or collected by the plant.

THE EFFECT OF FLOWRATE ON COLLECTION EFFICIENCY

Some types of dust collector are designed for a specific gas velocity and their performance can be adversely affected if operated at other gas velocities. Most collectors will accept flow reductions of around 20% without serious loss of efficiency, but still lower flows may impair the normal collection mechanism.

With high-inertia fibrous filters, some wet washers and, to a lesser extent, cyclones, the collection efficiency falls with reduced gas throughput, whereas for settling chambers and electrostatic precipitators the collection efficiency will increase. Bag filters and variable area scrubbers provide fairly constant collection efficiency from zero flow up to their full rated capacity.

FLUCTUATING TEMPERATURES

Temperature variations are most difficult to handle when the gas temperature varies around its dewpoint. Under these conditions fibrous and fabric filters are easily blinded and are then difficult to recover into normal operation. Condensation will generally result in the dust becoming sticky (see above).

EXPLOSIVE AND COMBUSTIBLE GASES

Gases which are within their limits of flammability must never be allowed to enter an energised electrostatic precipitator. The risk is also high in a fabric filter

43

if any form of ignition is possible. In some applications, dangerous waste gases can be rendered safe by incorporating a well-designed air injection system, followed by controlled combustion in a refractory-lined chamber. The resulting fully-oxidised gas mixture can then be cooled prior to treatment in a precipitator or fabric filter.

If there is no air present, fuel gases are commonly treated by electrostatic precipitators. If oxygen is present, dedusting should be limited to treatment by inertial separators and wet washers. Even then, great care is necessary to avoid all risk of spark ignition, both in the collector and in the ducting.

CORROSIVE GASES

All dry collectors must be designed and operated with care when corrosive conditions are present. Where possible, it is best to keep such collectors above their dewpoint by a combination of generous lagging and supplementary heaters. Cold spots, leading to corrosion, often occur in hot-gas plants due to inleakage of air around badly-fitting flanges and doors, etc. Particular care should be taken to avoid condensation occurring during start-up and shut-down.

Wet washers will accept corrosive conditions if correct materials of construction are used, but special attention is necessary at any wet/dry interface. Chemical dosing systems may be required, particularly on those wet washers which are fed from a recirculated water system.

ANCILLARY EQUIPMENT

As already noted in earlier sections of this guide, a dust collection system is rarely a single item of plant. Having removed the dust from the gas stream, it has to be brought down into a transfer hopper, extracted and possibly conditioned, thickened or otherwise treated.

Wet collectors will often require some form of water treatment plant, which may be many times more expensive and complex than the actual wet collector itself. Even dry collectors may require sophisticated mechanical or pneumatic dust handling systems, particularly if they are associated with dust involving more than, say, 50 kg per hour of solids. Thus, the complexity of ancillary equipment is more often related to the weight of dust collected rather than to the type of collector used.

ON-LINE REGENERATION

Those types of single-chamber fabric filters incorporating cleaning by mechanical shaking, and also intermittently-flushed electrostatic precipitators and ag-

44

gregate bed filters, have to be periodically isolated from the process gas stream in order to remove collected dust from the plant internals. Fixed element fibrous filters also need to be isolated for fitting replacement pads. Sometimes this can be accommodated within cyclic variations of the dust production, as in batch processes, but often this is not possible. The alternative is to install surplus filtering capacity and to divide the collector into a number of stages, one or more of which can be periodically isolated or de-energised for cleaning. Obviously such an arrangement is likely to increase capital cost and plant complexity.

Those types of plant which do not need to be so divided and which can operate continuously on the process gas stream in a fully operative mode include cyclones, wet washers, pulse-jet fabric filters and most dry or continuously irrigated forms of electrostatic precipitator.

8. CYCLONES AND INERTIAL SEPARATORS

Cyclones are mainly used either as pre-collectors or for primary product collection. The operating principle of these devices is that a spinning motion is imparted to the dusty gas. The particles are thrown outwards by the centrifugal force to the wall of the vessel and are swept down to be removed via a discharge hopper. Cleaned gas is removed from the central portion of the cyclone.

For the collection of coarser dusts, gravity and inertial separators are sometimes used instead of cyclones. These devices are employed almost entirely as precleaners. They basically consist of expansion chambers in which the gas velocity is reduced to such an extent that coarse particles can settle out under gravity. To improve collection, the gas may also be induced to undergo abrupt changes of direction by the use of baffles.

8.1 CYCLONES

DESIGN PRINCIPLES

The most popular type of cyclone is the reverse-flow cyclone, shown diagrammatically in Figure 8.1. The dust-laden gas enters the body tangentially and spirals downwards in a vortex motion. At the base of the unit the direction of axial flow reverses and the cleaned gas leaves the unit axially by spinning upwards in a tight, fast, central vortex. Particles, which have moved to the wall, pass down to the collection hopper where they are removed. Cyclones of this type are available with diameters in the range 0.1 to 2 m.

For a given dust, the diameter of the inner vortex is about half that of the gas outlet thimble. The inlet gas velocity determines the cyclone performance — the smaller the vortex and higher the velocity, the smaller the particles which can be collected. The cyclone capacity for a given pressure drop is also controlled by the vortex size and inlet velocity — again, a large capacity corresponds to a large diameter vortex. For the purpose of classification, reverse-flow cyclones can be described as either 'high gas throughput' or 'high efficiency' types. The former have a large outlet tube but small height-to-diameter ratio. They often incorporate a scroll inlet. For higher efficiencies, but

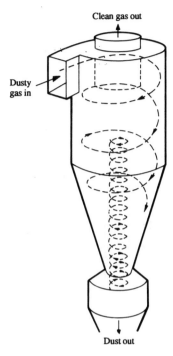

Figure 8.1   The reverse-flow cyclone.

lower throughput, the design alters to provide a vortex of smaller diameter but longer length. Figure 8.2 overleaf shows typical dimensions in terms of the diameter for the extremes of these types of cyclone.

In practice, most commercial cyclones do not provide ideal aerodynamics partly because the designs seek an optimum between throughput and acceptable efficiency. The main problem to overcome in the design of a cyclone is dust re-entrainment caused by particle bounce from the cell walls or excessive turbulence in the dust discharge hopper.

The inside surface of the cyclone should be as smooth as possible to maintain a stable, high-energy vortex and to reduce re-entrainment. The vortex finder or exit thimble of the unit is also important in stabilising the flow and minimising the bypassing of particles between the inlet and outlet; it should extend a little below the base of the inlet. The outlet pipe should be straight for two or three pipe diameters to avoid vortex instability, or alternatively an outlet scroll may be used to recover some of the spinning energy of the inner vortex.

47

The width of the inlet should be less than the gap between the cyclone wall and the outlet thimble. If the inlet is too deep, it reduces the effective length of the cyclone. A decrease in cyclone length increases dust carryover from the base but decreases cost. A length-to-diameter ratio of 2.5 to 4 is normally used. There are no advantages in fitting internal baffles or vanes. At the base of the unit, the outlet diameter should be larger than its vortex diameter, otherwise dust pickup and vortex instability will result. The cone angle should be about 15°. Since a negative pressure differential may exist across the hopper dust exit, a good seal must be maintained, otherwise back-flow, and hence re-entrainment, will occur. Rotary or double flap discharge valves are normally used to achieve dust discharge with minimum air ingress.

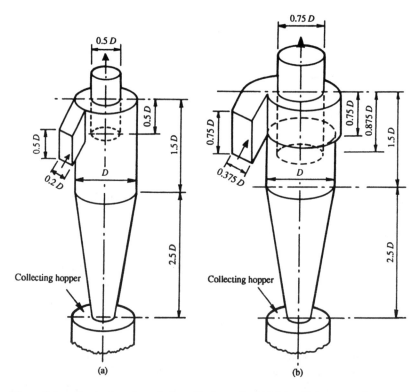

Figure 8.2   Standard cyclone designs (Stairmand); (a) high-efficiency, medium throughput pattern: normal flowrate = $1.5\,D^2$ m³/s; (b) medium-efficiency, high throughput pattern: normal flowrate = $4.5\,D^2$ m³/s. Entrance velocity at these flows is approximately 15 m³/s in both types. D in the above equations = diameter (m).

Where high gas rates have to be handled, it is convenient to arrange cyclones in parallel, multiple banks. With this arrangement, a higher overall collection efficiency and a lower pressure drop are achieved than with a single large cyclone. A common dust hopper can be used but care must be taken to ensure equitable air distribution. Division plates may be inserted in the hopper to prevent gas circulation between cyclones. Cyclones may also be arranged in series. Two cyclones in series operated at lower velocities may be used, for example, to collect fragile agglomerates which would break up in the high shearing fields normally generated in high-efficiency units. An example of this is the separation of the catalyst in a catalytic cracking process.

Small diameter cyclones which have a high efficiency but low individual throughput may be arranged in multiple banks in common housings for economy of construction. Figure 8.3 shows such a multi-cyclone. Good gas and dust distribution is necessary in this type of device to prevent backflow through individual cyclones, plugging, excessive and uneven wear and re-entrainment

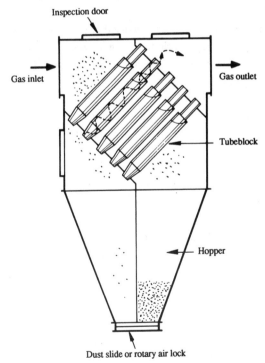

Figure 8.3   Multicell cyclone.

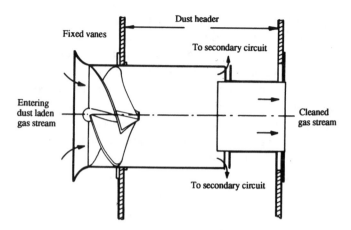

Figure 8.4   Fixed-impeller, straight-through cyclone.

from the dust bin. A small purge of about 5 to 10% is often used to stop backflow in the event of some of the tubes plugging.

An alternative type of unit which can either be used in multiple banks or as an individual unit is the straight-through cyclone shown in Figure 8.4. Here the spinning motion is caused by drawing the gas through a fixed propellor in a tube. The particles are thrown radially outwards and collected in an annular gap through which a small purge stream may be drawn. The efficiency of these units may fall if the particles are thrown outwards too violently so that they rebound inwards from the tube wall. These units are often used as primary filters in engine intake systems.

Cyclones may be irrigated by water injection either into the feed or into the cyclone body. The droplets are centrifuged outwards thus enhancing collection efficiency, and the irrigation of the wall reduces dust re-entrainment. Typical irrigation rates are 0.7 to 2 litres/m^3 of air. Entrainment of water droplets is normally minimised by the addition of a skid on the outlet thimble.

ADVANTAGES OF CYCLONES

* Low capital and running costs.
* Low maintenance.
* Moderate pressure drop.
* Minimum space requirement for a given gas throughput.
* Capable of operation with high dust loadings.

- Can be used with a wide range of gases and dusts.
- No moving parts.
- Units need no housing.
- Can be operated at high temperatures and pressures.
- Easy to keep hygienic.
- Manufacture possible in a wide range of materials.

DISADVANTAGES OF CYCLONES

- Low collection efficiency on small particles.
- Light materials or needle shaped materials difficult to remove.
- Plugging can result where dewpoints are encountered.
- Explosion relief for flammable materials is difficult.
- Potential problems with abrasive dusts.

CYCLONE PERFORMANCE

Apart from the dust particle size, the cyclone collection efficiency is dependent on a wide range of parameters including cyclone diameter, inlet velocity, particle density and gas properties. For reverse-flow cyclones, the d_{50} cut size, which is defined as the diameter of a particle which is collected with 50% efficiency, may range from about 2.5 μm to 20 μm. Typical grade efficiency curves, which show the efficiency with which particles of varying size are collected, are illustrated in Figure 8.5.

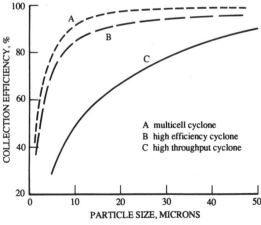

Figure 8.5   Typical grade efficiency curves for different designs of cyclone.

51

Since cyclones, like wet dedusters, are inertial separation devices, a high performance will result if the system inertial parameter, N_{st} (or Stokes' Number) is high. This is given by:

$$N_{st} = \frac{d^2 \rho\, u}{9\,\mu\, D}$$

where ρ = particle density
d = particle size
u = inlet velocity
μ = gas viscosity
D = cyclone diameter

Cyclones of the same geometrical design will have the same collection efficiency for the same inertial parameter. Hence, with certain restrictions with regard to velocity effects which are dealt with below, the cut size for a given design of cyclone is inversely proportional to the square root of the inlet velocity and the particle density, and proportional to the square root of the gas viscosity and the cyclone diameter. A number of mathematical models of cyclones are available in the literature[32-34].

The efficiency of dust removal increases with gas flow rate (due to the increase in Stokes Number) until a stage is reached where excessive turbulence is induced in the cyclone (see Figure 8.6). The flat maximum in this curve defines the normal operating range of the cyclone. The fall in efficiency which occurs at inlet velocities in excess of 30 m/s is caused by turbulence which gives rise

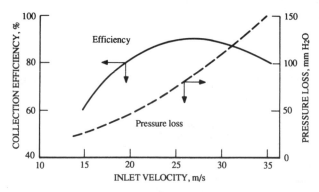

Figure 8.6   Typical cyclone collection efficiency and pressure loss as a function of inlet velocity.

to bypassing and particle re-entrainment within the cyclone. Where a large reduction in gas flow rate is anticipated, the drop in efficiency which would result can be avoided by arranging a number of units in multiple groups. At low flowrates some of the cyclones can be blanked off with dampers, thus maintaining a high inlet in those still in operation.

The pressure drop through a cyclone which is also plotted in Figure 8.6 is proportional to the square of the gas throughput.

Cyclones are particularly useful for handling high dust loads since the efficiency increases and pressure drop falls with increase in loading. Care needs to be taken when handling abrasive dusts, however, because wear can develop very quickly if the cyclone material is not suitably selected.

SPECIAL SAFETY FEATURES

Explosion relief vents must be provided when processing explosive materials and units should be earthed to avoid electrostatic charge build-up. Care must be taken to avoid leakage when handling toxic materials. The outlet thimble is supported by the roof and this should be taken into account in its mechanical design.

To be effective and safe, explosion vents need to be correctly sized, positioned and designed[10]. In addition, they must be vented to a safe area away from the operators. Explosion relief vents, which are located on the roof of the cyclone, are used when processing explosive dusts, but with Class A dusts it is extremely difficult to achieve the relief area without seriously affecting performance.

NORMAL MAINTENANCE REQUIREMENTS

Since there are no moving parts, extensive maintenance is not normally required. However, for operation in extreme conditions, access should be provided for periodic inspection. In the event of corrosion or erosion, replacement of linings or the entire unit may be required.

Wear frequently takes place at the base of the cone section immediately above the collecting hopper; it is good practice, therefore, on many applications to provide the cyclone with a short flanged section at the base of the cone to allow for easy replacement.

The cyclone may have to be taken off line for maintenance. If a number of units are used in parallel this will not necessitate plant shutdown. An isolation slide valve located above the rotary discharge valve facilitates its maintenance.

POSSIBLE PROBLEM AREAS

An occasional check of pressure drop and flowrate should ensure that the unit is operating according to design. Changes in performance may indicate internal plugging or gas leakage, but if this is not verified by internal inspection an alteration in the inlet dust load or size distribution should be suspected.

8.2 GRAVITY AND INERTIAL SEPARATORS

Devices of this type are used as pre-cleaners, usually with the objective of removing large, dense particles. Some typical designs are shown in Figure 8.7 opposite.

GRAVITY SETTLERS

Gravity settlers operate by allowing the effluent gas to expand into a large chamber. This reduces the gas velocity and causes the particles to settle out. The important factors in settling chamber designs are the surface area available for sedimentation, the terminal settling speed of the particles and the gas flowrate.

BAFFLE CHAMBERS

Another type of inertial separating system causes the gas to change direction by means of baffles. Both the baffle chamber and the settling chamber are seldom used in modern gas cleaning practice because they require a large floor area. In almost every case, a cyclone would provide better collection efficiency.

LOUVRES

An extension of the baffle principle, which increases the inertial separation of particles, is the use of banks of small baffles which split the gas flow as well as changing its direction. These baffles, which are normally referred to as louvres or chevrons, are occasionally used for dust control, but much more commonly to catch large droplets carried over from cooling towers or wet scrubbers and for demisting duties generally.

SKIMMERS

There are several devices available which use centrifugal force to achieve separation but which do not have the spiralling vortex characteristics of the cyclone. In a skimmer, the dust is carried by centrifugal force to the outside wall of the scroll from which the concentrated dust layer is skimmed off.

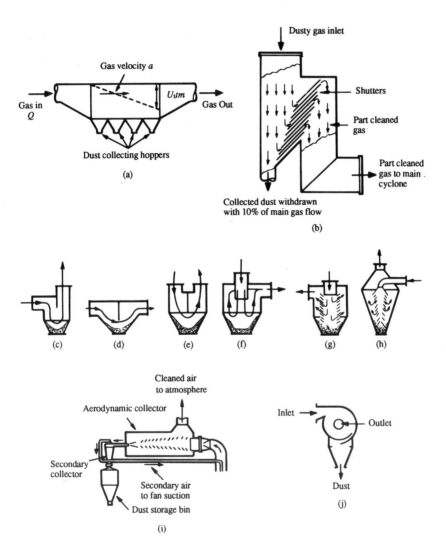

Figure 8.7   Gravity and inertial separators: (a) gravity settling chamber; (b) settling chamber with plates; (c, d, e, f) baffle chambers; (g, h, i) louvre chambers; (j) centrifugal skimmer.

MECHANICALLY-ASSISTED SEPARATORS

These contain an impeller to provide the centrifugal force which cleans the gas and discharges the dust. On entry, the gas is turned through a right angle and spun by the impeller against a curved back plate which concentrates the dust so that it can be discharged through an annulus into the hopper. A device of this nature can achieve efficiencies as high as those of the cyclone, but the running costs of the impeller have to be taken into consideration.

9. WET DEDUSTERS AND DEMISTERS

The characteristic of wet dedusters which sets them apart from other gas cleaning equipment is the use of a scrubbing liquid, normally water, to achieve collection of the particulate material. The scrubbing liquid is usually dispersed in a spray, or spread in a film over the internal surfaces of the scrubber. Downstream of the collection zone some form of spray eliminator is normally required. The equipment and mode of operation used in demisters, which collect fine liquid particulates, are essentially similar to those of dedusters.

The major mechanism by which fine particles are wetted in a deduster is by impaction on to wet surfaces, most commonly droplets. As the gas and particles flow around a droplet, the tendency of a particle to be captured depends upon the balance which is set up between its inertia, which causes it to move towards the drop, and the aerodynamic drag force exerted upon it by the gas, which tends to cause it to move around the droplet and escape capture. Thus large, fast-moving particles are collected more readily than small particles moving at lower velocities. Other mechanisms are sometimes significant in wet washers. The most notable of these is condensation, which can increase the effective diameter of the particles, thus making them easier to collect.

Most wet washers may also be used for gas absorption, for example, sulphur dioxide removal. This type of application will not be considered in this guide.

9.1   ADVANTAGES AND DISADVANTAGES OF WET COLLECTION

There are many advantages and disadvantages in wet collection. These are summarised below. In particular, it can be noted that wet collection is advantageous when good efficiencies are required chiefly on coarse dusts, for low loadings where only moderate efficiencies are required or in situations where there is a significant fire or explosion risk. The principal disadvantages of wet scrubbers are the requirement for disposal of the collected material in a wet form, the low efficiencies on small particles (except at very high pressure drops) and the loss of thermal buoyancy and resultant increased visibility of the stack plume.

ADVANTAGES

• Compactness and hence low capital cost.

• Elimination of fire and explosion hazards.

• Possible neutralisation of corrosive material by the selection of an appropriate scrubbing liquid.

• Simultaneous removal of gaseous and particulate contaminants.

• Simplicity of operation.

• Lack of secondary dust problems in disposal.

• Material is recovered in a form which can be pumped.

• Efficiency is largely independent of scale.

• Continuously rated at a constant pressure drop.

• Possibility of cooling and dedusting in one item of plant.

• Can be used to collect sticky dusts.

• Insensitive to changes in gas temperature near to the dewpoint.

• Potential for low-grade heat recovery system.

DISADVANTAGES

• Requirement for a slurry treatment plant.

• Extra cost of water usage.

• State of material is changed, ie it is collected wet.

• Loss of thermal buoyancy in the stack plume.

• Increased visibility of the stack plume due to water vapour condensation.

• Efficiencies do not approach 100% and therefore:

(i) cleaned air cannot be recirculated into working areas;

(ii) not necessarily suitable for high inlet burdens;

(iii) the fine fraction, if not collected, can make the plume visible;

(iv) high pressure drops and hence high power consumption is necessary for operation at high efficiency or on sub-micron particles.

• Frequent inspection and regular maintenance usually required.

• Possibility of scrubbing liquid freezing, foaming, frothing or gelling.

• Possibility of retarded settling of the collected material.

• Possibility of adverse chemical reactions with the scrubbing liquid which might lead to corrosion or deposition problems.

9.2   TYPES OF WET COLLECTORS

Most types of wet deduster and demister tend to be hybrids in that they use more than one approach to the problem within the same unit. In general, wet collection equipment may be divided into groups as follows:

- baffle spray eliminators;
- cyclonic spray eliminators and dedusters;
- packed-bed demisters and dedusters;
- fibrous demisters and dedusters;
- mechanically-assisted dedusters;
- plate scrubbers;
- self-induced spray dedusters;
- venturi and orifice scrubbers;
- electrically augmented scrubbers;
- foam scrubbers;
- gravity spray scrubbers.

The first four groups are more often used for demisting than for dust removal. They can, however, cope with soluble particles or low concentrations of solids. There are many design variations within each class and Figure 9.1 overleaf gives schematic diagrams of some representative devices within each group.

Baffles are most often used for the removal of large diameter (50 μm) droplets. They are placed so as to cause the gas to change direction thus inducing inertial impaction. One of the commonest arrangements is the zigzag baffle or chevron (Figure 9.1(a)). Straight baffles or louvres (Figure 9.1(b)) are also used and although these are generally operated at lower approach velocities they are easier to maintain. There are many design variations on this simple theme which aim either to reduce pressure drop by using streamlined baffles, or to reduce re-entrainment and so allow an increased operating approach velocity. Typical approach velocities for ordinary zigzag baffles are in the region of 2.5 m/s.

Centrifugal devices are used both as demisters (Figure 9.1(j)) and, in conjunction with sprays, as wet dedusters (Figure 9.1(c)). Their capabilities are the same as those of the cyclones discussed previously. A serrated skin, fitted to the exit thimble, is normally used to prevent short-circuiting of the liquid under the influence of the annular eddy.

Irrigated packed beds are widely used for mass transfer and they may also be used for the removal of particles. Since high dust loadings can cause

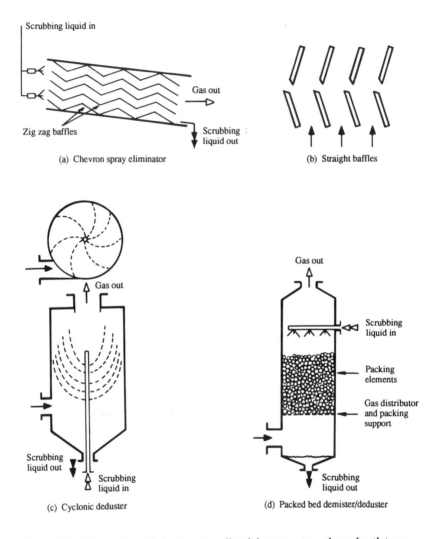

Figure 9.1　Types of wet deduster. ▶ = dirty inlet gas; ▷ = cleaned outlet gas; ▷▷ = scrubbing liquid inlet; and ▶▶ = scrubbing liquid outlet.

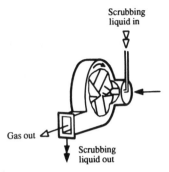

(e) Mechanically assisted deduster

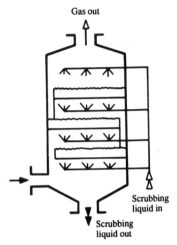

(f) Plate scrubber

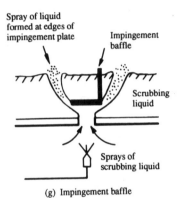

(g) Impingement baffle

(h) Bubble cap

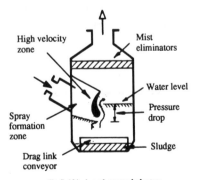

(i) Self induced spray deduster

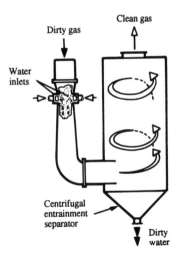

(j) Venturi scrubber

packed towers to clog, they are used mainly for demisting and the removal of soluble solids. The aerosol is usually passed up through a bed of packing countercurrent to the scrubbing liquid which is sprayed onto the bed surface (Figure 9.1(d)). In tall beds some means of ensuring even liquid distribution should be employed. Many of the standard, commercially available packings used for mass transfer operations may be employed as well as spheres or crushed rocks. For certain specific applications where there are known problems with scale deposits, mobile bed scrubbers have been successfully used. In these, a bed of spheres is fluidised by the gas flow whilst the scrubbing liquid flows downwards countercurrently. The continuous movement prevents the formation of deposits whilst also providing an active mass transfer surface.

Fibrous demisters, incorporating pads of uniformly crimped wire (normally stainless steel), are used for the removal of large diameter droplets. For fine mists, the pads are formed of smaller, randomly packed fibres (often hydrophobic). They may be used for the collection of low concentration dusts if they are continuously irrigated. The randomly packed fibre pads are analogous to fibrous filters and if operated at a low face velocity can achieve high efficiencies on very fine mists. In an alternative arrangement, the fibrous pads are built into a slowly revolving drum, part of which dips into the liquor reservoir so that the pads are continuously wetted.

Mechanically-assisted scrubbers incorporate some form of mechanically-driven rotor within the wet collection zone. The rotor can be used either to draw the gas through the unit, and so provide a wetted surface for impaction (Figure 9.1(e)), or it may be used to disintegrate the liquid to a fine spray into which the particles may impact. Whilst this last category can achieve high efficiencies, its use has declined in recent years because of its high operating costs.

Plate scrubbers are towers in which the aerosol is bubbled through one or more liquid layers in series. Each layer is supported upon plates within the tower. Flow is countercurrent and sprays are often installed underneath the plates to prevent the formation of wet/dry interfaces (Figure 9.1(f)). There are three common types of support for the liquid layers. The first is a simple perforated plate for which particle capture occurs by inertial impaction during the bubble formation process; the second is a bubble cap plate and the third is the impingement baffle plate, depicted in Figure 9.1(g). The majority of the particle collection for this type of plate occurs in the spray formed in the high velocity zone at the edges of the impingement baffle.

By far the most important mode of wet collection is onto sprays of droplets which are produced within the scrubber by nozzles or more commonly by self-induction caused by the dusty gas flowing at high velocity across the surface of the liquid. Spraying nozzles for wet scrubbers tend to be one of two major types, referred to as pressure nozzles and two-fluid nozzles. Rotating discs or sonic devices may also be used.

In self-induced spray scrubbers (Figure 9.1(i)), the dirty inlet gas is accelerated through a constriction, one wall of which is formed by the scrubbing liquid. The highly turbulent gas flow atomises the liquid into a fine spray which initially moves very slowly. The resulting large relative velocity between the particles and the spray droplets promotes efficient inertial impaction. This, together with the centrifugal force caused by the rapid changes in the direction of the flow, ensures that efficient overall collection occurs. No moving parts are required. These units are normally sold complete with sedimentation tank and slurry removal gear.

The venturi scrubber (Figure 9.1(j)) in its simplest form consists of a constriction in the duct carrying the dust-laden gas which raises the velocity to between 60 and 90 m/s or even higher. Scrubbing liquid is introduced at or upstream of the constriction or 'throat'. The high gas velocity atomises the liquid and the high relative velocity between the accelerating liquid droplets and the dust particles in the gas leads to the very efficient collection of even the fine particles. The liquid droplets are then separated from the gas stream in the cyclonic separator. Sometimes, the scrubbing liquid is injected upstream of the venturi through high pressure spray nozzles. Most venturis have provision for adjusting the throat area and hence the operating pressure drop. The venturi scrubber is the most efficient type of scrubber, except perhaps at low pressure drops; it can be operated at very high pressure drops in order to achieve high collection efficiencies. In recent years, venturi scrubbers have replaced many of the less efficient types of scrubbers.

In gravity spray scrubbers, liquid is sprayed downwards into a tower from one or more banks of spray nozzles, generally located at or near the top, whilst the dust-laden gas enters the bottom of the tower and flows upwards. To avoid droplet entrainment into the exit gas stream, the terminal settling velocity of the droplets must be greater than the upward velocity of the gas, hence limiting the smallest size of droplet that can be used. The spray tower probably represents the simplest type of scrubber and is often used for treating large volumes of gas (100 m^3/s) because of its low pressure drop and the minimal risk of blockage. It can also be used as a precooler or as an absorber for acid gas removal where solids deposition would be a risk in other types of scrubbers.

9.3 RECENT DEVELOPMENTS IN WET COLLECTORS

Several recent developments have aimed to reduce the high power dependence of wet dedusters when dealing with fine particulates. One such technique has been the incorporation of a high-voltage ionising field within the scrubber to provide an electrostatic charge to the dust or to the water droplets, or both. Electrically-augmented scrubbers are available in several forms and there is growing support for the claim that ionising power is more energy efficient than fan power for improving the collection efficiency of sub-micron dusts. A comprehensive review of the subject is given in Reference 41.

Use of a 2- or 3-stage venturi scrubber instead of a conventional single-stage unit has been shown to improve scrubber performance significantly[42,43]. This results from the fact that for a given particle size, a 'critical' pressure drop is reached above which the particle scrubbing process becomes less energy-efficient and it then pays to split the operation into more stages. The critical pressure drop increases as the particles become finer. Both theory and experiment indicate that optimum performance is achieved with a 3-stage venturi.

Another method of improving scrubber performance uses the bubbles in a stable bed of foam as collecting surfaces. As the dust-laden gas passes through the bed, the dust particles impact upon the bubbles, some of which break down into liquid which runs countercurrent to the gas to the sludge outlet at the base of the unit. As the foam has to be constantly replenished, the economics of this type of unit are strongly dependent upon the amount of surfactant used.

Flux force and condensation scrubbers represent another approach to improving scrubber performance. In this case, a hot humid gas is brought into contact with a cold liquid. This is claimed to promote a certain amount of diffusiophoretic (flux force) collection in addition to inducing condensation on to the dust particles, thereby increasing their effective size. To be economically feasible, waste heat, preferably as steam, must be available.

There are, however, few industrial installations of any of the above types of scrubbers from which reliable plant performance data are readily available.

9.4 PERFORMANCE OF WET COLLECTORS

The efficiency of wet scrubbers is not only dependent upon the particle diameter but also upon the particle density, the liquid-to-gas ratio and the relative velocity of the aerosol and droplet in the collection zone (and hence the energy consumption). 'Typical' grade efficiency curves are thus only an approximate guide to the actual performance that may be expected from a unit operating at an optimum level. Figure 9.2 gives examples of some typical grade efficiency curves. There are many mathematical models of scrubbers available in the literature[36–38].

In principle, it would appear that it is difficult to make general statements upon equipment performance because the designs of individual scrubbers vary so widely. However, it should be pointed out that collection generally will be by inertial impaction and the fine particles which pass through one stage may well also pass through each subsequent section. Thus, the overall efficiency is usually limited by the maximum value of the inertia parameter which can be obtained at any stage within the unit (see Chapter 8).

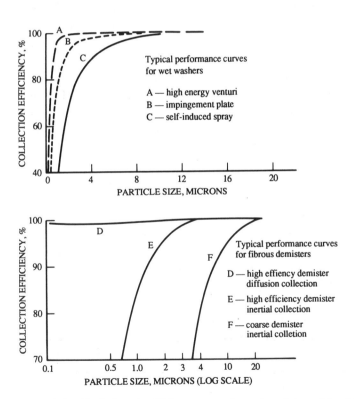

Figure 9.2   Typical grade efficiency curves for wet washers and demisters.

65

Experience has therefore shown that the efficiency of wet dedusters is strongly dependent upon their total energy usage, particularly upon the pressure drop across the collection zone. Moreover, provided the liquid is evenly dispersed within the scrubber, similar pressure drops will often give similar efficiencies on the same dust for quite different scrubber configurations. Thus, it is not possible to design a wet deduster which has high efficiencies on particles of 1 μm and below, and which does not have a high energy consumption and hence high running costs. When comparing wet scrubbers, therefore, the purchaser should bear in mind that he will only achieve the efficiency that he is prepared to pay for. Prospective purchasers are advised to treat with caution any claims to the contrary.

As a result of this dependence of performance upon pressure drop, attempts are often made to classify scrubbers according to their pressure drops. Low-, medium- and high-energy scrubbers are therefore frequently referred to by the approximate and arbitrary dividing lines between the categories of < 100 mm water gauge pressure loss for low-energy scrubbers, and > 500 mm water gauge pressure loss for high-energy scrubbers.

In principle, if the gas flow rate through a wet deduster varies significantly then so does the pressure drop across the unit and hence the efficiency. For self-induced sprays, a decreased velocity results in a larger droplet size and a decreased relative velocity between droplet and particle. Thus, inertial collection is reduced. Consequently, many commercial designs of self-induced spray and venturi scrubbers provide a means of controlling the pressure drop across the unit to enable a constant efficiency to be maintained in spite of changes in throughput. For those scrubbers which incorporate plates, packings or spray nozzles this generalisation is valid up to the throughput at which flooding occurs. However, at flowrates above this the unit may still act effectively as an agglomerator, with the entrained mist being of a larger diameter than the inlet dust and thus easier to separate. For those units where the relative velocity between the particle and collector is caused by the rapid motion of the collector (eg mechanically-assisted scrubbers or devices incorporating high speed sprays) an increased throughput can result in a decrease in efficiency.

Recently, some of the previously accepted applications of wet scrubbers have come under close scrutiny because of their inability to cope with modern, increasingly stringent, emission requirements. Moreover, while high-energy scrubbers are capable of maintaining good efficiencies down to particles of 0.5 μm and below, their high running costs may give a strong motivation to find economic alternatives. However, wet scrubbers often have low capital cost, and in some situations, the collection of a dust in a wet state is desirable.

9.5   SCRUBBING LIQUID TREATMENT

In most wet dedusters the scrubbing liquid, usually water, is clarified to a sufficient degree to enable it to be reused. The clarification equipment is an integral part of the dedusting system and should be carefully considered from the outset. Unless surplus effluent treatment capacity already exists on a plant it will add significantly to the cost of the scrubbing equipment (indeed, in some instances it may cost more than the collector itself).

The closed liquid cycle has the advantages of lowering the overall water consumption, of minimising the solid/liquid separation equipment required and of ensuring a sufficiently high liquid-to-gas ratio. This last parameter is important in the operation of wet dedusters. It is defined as the ratio of liquid dispersed in the collection zone to the gas throughput. Wet dedusters usually operate with a ratio in the range 0.1 to 10 m^3 of liquid per 1000 m^3 of gas. If the liquid-to-gas ratio is too small there will not be sufficient distribution of liquid within the collection zone and efficiency will be impaired. If too high a value is used then water recirculation rates become excessive with little or no increase in efficiency.

The scrubbing liquid regeneration system must also maintain the concentration of suspended solids in the recycled liquor at an acceptable level. Commonly, concentrations of up to 10% are tolerated, but these solids contribute to erosive wear on impingement surfaces in the recycle pump and in any spray nozzles. Wire strainers or screens should be used to protect pumps and sprays from damage or blockages by large pieces of debris in circulating liquors.

Even when the scrubbing liquid is re-used it is still necessary to add fresh make-up water to the system to compensate for water lost with the sludge, and also for losses due to evaporation. Some form of liquid level controller is necessary in most scrubbers to keep the liquid hold-up in the unit constant. With hot gases, large quantities of make-up water can be needed to compensate for evaporation losses. It is normal practice to add make-up water into the clarified water tank. A single pump is then used to feed both the pre-cooling sprays and the main scrubber sprays. In very high efficiency applications, however, it is advisable to use the cleanest water in the upper sections of the washer.

Where a scrubber liquor is operating on gases containing acidic sulphur or chloride compounds it will almost certainly dissolve a fraction of these constituents, irrespective of whether this was intended in the design or not. Even with hot, non-acidic gases the high evaporation loss can magnify the effect of the chloride content in the make-up water and this itself can be a major source of the acid radical.

Such acidic ions, being soluble in the wash liquor, are not preferentially removed in the thickener. There is a risk, therefore, that with continuous recycling of the liquor the concentration of even minor constituents can build up to corrosive levels. As even low concentrations of acid can be harmful to mild steel plant, it is vital to compare the rate of accumulation of these components with the rate of discharge in the under-flow thickened liquor. If a build-up is likely to result, additional means of purging the contaminant must be found. Ways of achieving this include neutralising its effect chemically, converting the material to an insoluble form that would be removed with the slurry or bleeding more wash liquor from the recycle circuit; pH control and alkali dosing are commonly used.

The correct choice of instrumentation and control equipment is important if the protection system is to achieve its desired effect. pH sensors are best located in the clarified liquor stream where the effects of entrained dust particles are minimised.

Another common source of problems in water treatment plants occurs when calcium salts are prevalent. In the presence of carbon dioxide, calcium carbonate deposits can be formed and these are notoriously difficult to remove once formed. Control of the deposit is possible with a scale inhibitor but expert advice on application is recommended.

In contrast, polymeric flocculant aids can be used to increase deposition in the clarification plant, especially when a fine dust is being collected. In some cases, both flocculants and anti-fouling agents can be used in the same scrubbing system, but particularly careful design is necessary in these circumstances.

Foaming can also be a nuisance. It is made worse by high concentrations of fine particles and dissolved surface-active solids. Specialised anti-foaming agents can be effective but care is required in their application to avoid their causing difficulties. They might, for example, affect the performance of a froth flotation cell or sedimentation tank in an adjacent part of the liquor circuit.

The dosing of all of these chemicals into the scrubbing liquor system generally requires fairly precise monitoring; they cannot be added indiscriminately. A reliable and accurate pump, normally of either the diaphragm or piston type, should be used. Occasionally, mixing vessels are incorporated into the system to ensure adequate dispersal of the dosing agents.

Water treatment plants for wet scrubbers are discussed further in Chapter 13.

9.6 NORMAL MAINTENANCE REQUIREMENTS

Correct maintenance of wet scrubbers is best achieved according to the maxim 'little and often'. Particular attention should be paid to the cleanliness and condition of the inlet headers and sprays, the spray eliminator(s), the fan impeller and the mechanical components of any drag-link conveyors.

In general, wet dedusters should be designed so that inspection panels or doors allow access to all points which can wear, clog or otherwise give trouble. Adequate ladders and platforms should be integrated into the support structure.

If, during normal operation, the air entrained into the collector is below the freezing point of the scrubbing liquid, antifreeze should be added to prevent freezing. (The antifreeze should be drained from the tank when weather conditions improve.) Electrical heaters are normally only capable of making up the heat lost from the tank walls in static conditions.

10. ELECTROSTATIC PRECIPITATORS

The electrostatic precipitator is capable of operating over a wide range of conditions of temperature, pressure and dust burden. It is not particularly sensitive to particle size, and can collect dust in both wet and dry conditions. Corrosion and abrasion resistance can be accounted for in the design.

The pressure drop across the precipitator is very low and the consequent savings in operating cost (fan power) may counteract its relatively high capital costs. This balance between operating and capital costs is more critical in larger installations and therefore precipitators are generally not considered for applications with gas flowrates below about 10 m³/s, although small units are used in conjunction with ventilation systems.

10.1 PRINCIPLE OF OPERATION

The removal of particulates from a gas stream by an electrostatic precipitator depends on the fact that charged particles in an electrostatic field will migrate towards regions of opposite polarity. The practical realisation of this basic law of physics is achieved by directing the gas through a number of earthed channels in which high voltage discharge electrodes are centrally positioned at regular intervals.

Figure 10.1 illustrates the process. Uncharged particles entering the channels receive a charge (usually negative) from the electrodes or ionised gas and then move towards the walls of the channel which are earthed. For effective dust removal the voltage between the discharge electrode and collector plates must be sufficient to achieve a stable corona discharge (the condition when the molecules of the gas are themselves ionised and cause a steady current flow) but insufficient to cause excessive sparking. The term electrostatic is misleading as a current is always flowing from the discharge electrode, through the gas space and down to earth. The current is carried mainly by gaseous free electrons and ions; because they exhibit a much lower mobility, the proportion of current carried by charged particles is low.

The required treatment time in an electrostatic precipitator, and conse-

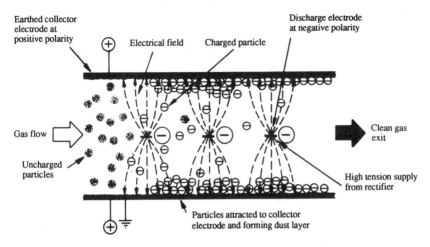

Figure 10.1   Principle of electrostatic precipitator operation.

quently its size, depends directly on the required degree of cleansing of the gas and the physical properties ('precipitability') of the dust.

Particles of the dust attracted to the earthed collector sheets form an agglomerated layer on the surface which must be periodically removed. The two principal methods are dry rapping, striking the surface of the collectors with a hammer device, or continuous or periodic flushing with a liquid (normally water) which usually results in a slurry effluent.

Whilst most precipitators operate with dry rapping, the wet precipitator is useful for removing dust from wet gases, or gases with temperatures close to the dewpoint, and for collecting liquid aerosols such as acid mist or tar.

10.2   TYPES AVAILABLE AND GENERAL DESCRIPTION

10.2.1   DRY-TYPE PRECIPITATORS

Figure 10.2 overleaf shows the general arrangement of a typical parallel-plate precipitator which is used for catching free-flowing dry dust from relatively dry gas (a gas at a temperature significantly higher than its dewpoint). The dust particles collecting on the surface of the collector plates are periodically removed by rapping or shaking of the plates and fall into the hoppers below. The dust collected in the hoppers is dry and usually convenient for handling and

71

disposal. The collected dust may be returned upstream into the process (eg as sinter strands), sold as a by-product (power station fly-ash for aggregate manufacture) or sold as a product (cement).

The dry type of precipitator is preferable for cleaning gases which will be discharged to the atmosphere, as the thermal buoyancy of the effluent gas will be unaffected and the possibility of forming a steam plume is kept to a minimum.

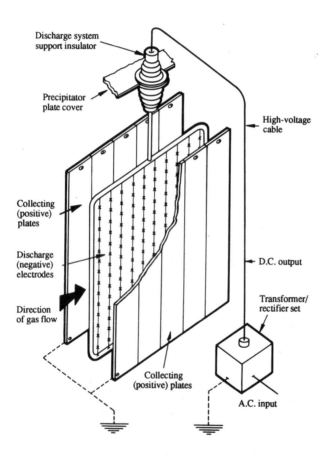

Figure 10.2   Essential features of the electrostatic precipitator.

Important characteristics of the dry precipitator are as follows.

COLLECTOR ELECTRODES

These are generally made of mild steel for cheapness, but other materials, particularly stainless steel, are used where corrosion or abrasion resistance is required. Single collector electrode plates are between 5 m and 15 m high. To provide the required rigidity with thin section material, the plates are either interlocked or incorporate shallow section ribs. Welding is rarely used as this can warp the plates. Every collector plate must be straight and rigid and close tolerances are essential in manufacture and erection.

Systems for supporting the collector electrodes in position must not only provide support and accurate positioning but must also allow transmission of the rapping blow over the whole surface to ensure effective dust removal. A dust build-up at only one point can severely reduce the effectiveness of a complete section of the precipitator and may cause it to short out completely.

DISCHARGE ELECTRODES

Many electrode types are available such as twisted wire, barbed wire and saw tooth. Some designs hang freely, weighted at the lower end to ensure they hang vertically, whilst other designs are rigidly held in a frame. The different discharge electrode designs exhibit different corona characteristics and selection is usually based on experience with a particular process.

The mechanical reliability of the electrodes and their supporting frame is important as a single broken wire can short out an entire electrical field of the precipitator. The discharge electrodes must also be rapped or vibrated to prevent dust build-up and their mechanical strength must be compatible with transmission of the rapping blow or vibration.

The discharge electrodes must be particularly resistive to any corrosive elements in the gas stream.

COLLECTOR ELECTRODE RAPPING

The collector plates are generally rapped by tumbling hammers (internal or external) or by a drop-rod mechanism, driven by rotary drives or sometimes by vibrators or solenoids. The rapping force must be sufficient to dislodge dust from the entire collector surface, but insufficient to cause excessive re-entrainment of dust into the gas stream. The rapping cycle is generally staggered throughout the precipitator to minimise the effect of re-entrainment. Collectors in the outlet section of the precipitator, where least dust is caught, are rapped much less frequently than those in the inlet region.

DISCHARGE ELECTRODE RAPPING

The discharge electrodes must also be rapped to remove any dust that collects. This is generally easier than collector rapping, although the supporting insulators must withstand the vibrational shock or a design must be used which protects the insulators from this shock. The electrode rapping system must be insulated as the rapping hammer will be periodically in contact with the discharge electrode high-voltage supply. Alternatively, the appropriate electrical sub-section can be turned off during rapping, but this will cause a temporary adverse effect on performance.

10.2.2   WET-TYPE PRECIPITATORS

In this type of precipitator, the collected dust is removed from the collector's plates by flushing with a suitable liquid, usually water, either intermittently or by continuous spray irrigation. For a few applications, such as acid mist precipitation, no further irrigation is required other than that provided by the collected mist itself.

Whilst in general dry precipitation is preferable, wet precipitators are useful for removing dust from wet gases or gases with temperatures close to the dewpoint and for collecting liquid aerosols such as acid mists and tar. The performance of wet precipitators is less dependent on particle properties as the moisture present in the gas precipitates readily and will assist the precipitation of a difficult dust. Re-entrainment is minimal and so wet precipitators are sometimes used for meeting very low emission requirements, including cleaning gases such as blast furnace gas prior to combustion. The wet precipitator effluent gas will generally produce a steam plume when discharged into the atmosphere.

The internals of the wet precipitator are principally the same as for the dry type with the following exceptions:

• The spray or weir system is designed to remove efficiently all of the collected material from the collector plates and to drain into the hoppers. It may be necessary to de-energise the field whilst it is being flushed.

• The materials of construction of the internals must be compatible with more severe corrosive conditions than in the dry precipitator.

• The precipitator produces a liquid or slurry effluent which will require a water treatment plant (see Chapter 13).

• Precautions must be taken to remove any water particles which might otherwise be carried out of the precipitator in the gas stream. Demisters may be installed or sometimes the last electrical field in the precipitator can be operated dry.

Wet precipitator designs are more often based on tubular collecting electrodes than those for dry precipitators. Tubular units are more suitable for de-tarring applications as it is easier to make them explosion resistant. However, tubular units generally have only a single electrical field and are less suitable for achieving very high efficiencies. It is feasible, however, to install plate-type wet precipitators inside cylindrical pressure vessel casings and this is often done for blast furnace gas applications.

10.2.3   COMMON FEATURES OF WET AND DRY PRECIPITATORS

TRANSFORMER-RECTIFIER SET
This converts the available low-voltage, alternating-current supply to the high-voltage, direct-current supply required by the precipitator. There are two principal types of rectifier in modern use, the selenium and silicon rectifier. These have largely superceded the old mechanical and thermionic valve rectifiers.

CONTROL SYSTEM
The high-tension (HT) voltage/current characteristics of an electrostatic precipitator are extremely sensitive to changes in process conditions such as temperatures, humidity or dust burden. It is therefore essential to have an automatic control sytem which will always maintain the HT power supply at its optimum level in order to achieve the best performance from the precipitator. A control system based on a preset primary voltage or current is unlikely to achieve maximum dust collection efficiencies, unless the dust producing process is very stable.

The highest efficiency of a precipitator is achieved when the maximum electric field strength is consistently imposed on the electrode system. In practice, this means that the voltage between the discharge and the collecting electrodes of the precipitator should be maintained at the maximum possible value. However, the current passing between the electrodes is a combination of corona discharge and energy contained in arcs or sparkovers and as such represents a very unusual electrical load. If the primary voltage into the transformer is gradually increased as in Figure 10.3 overleaf, the HT voltage also increases until a maximum is achieved. Beyond this point, further increasing the primary voltage

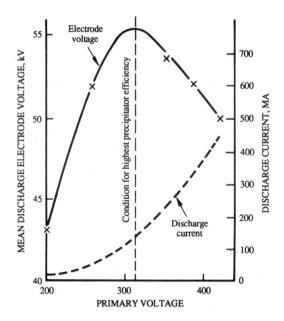

Figure 10.3 Typical electrical characteristics of an electrostatic precipitator.

also increases the sparkover and the HT voltage falls. The automatic control system must continually seek out and maintain this peak HT voltage condition.

The two main types of control system most frequently used are as follows:

(i) The spark-counting system uses a digital counter to monitor and maintain the number of sparkovers occurring within a specified range. If the sparkover rate goes outside the limit of the operating range the primary voltage will be adjusted accordingly.

(ii) The 'hill-climbing' technique uses step increases in primary voltage with a 'sense' circuit to detect whether the HT electrode voltage has increased or decreased. As long as the HT voltage increases, the step increases in primary voltage continue. When a reduction in HT voltage is sensed then the controller reduces the primary voltage by two steps, bringing the operating point back to the region of maximum electrode potential. If no process changes occur the controller will continue to oscillate around this maximum point.

Both control systems make their adjustments on the primary voltage using either transductors or, increasingly, thyristors. Thyristors have a much faster response, but the robustness of transductors can still be valuable.

The sensitivity of the control system has to be adjusted according to the process operation. Many processes operate continuously with very few sudden changes, eg boilers or cement plants, whereas other processes, such as many steel-making, are subject to almost continuous changes.

GENERAL LAYOUT AND SUB-SECTIONING

Figure 10.4 overleaf illustrates the general layout of a precipitator. The division of the unit into two or more completely separate chambers allows for one unit to be completely off-line and isolated for internal inspection and maintenance whilst the total gas flow, or a reduced gas flow, is routed through the remaining chamber(s).

Certain design features, such as the weight of an electrode suspension or length of a rapping drive, may give an upper limit to the width of a single chamber. The number of chambers, however, is usually a compromise between maintenance requirements and capital expenditure, and with the improving reliability of electrostatic precipitators single chamber units are often adequate.

The sub-division of the unit into several separate electrical 'fields' is advantageous as each field operates on a different dust loading (most dust is removed in the first field) and the electrical characteristics can be quite different. Each field can therefore operate independently with a separately optimised HT electrical supply. The number of fields varies from one to five (or more), the choice generally depending on how clean the effluent is required to be.

It is generally found that precipitator performance falls off if the length of discharge electrode wire supplied by a single transformer/rectifier set is too long. Therefore, particularly for large precipitators, there may be further sub-sectioning of the electrical supply. With a reasonable degree of sub-sectioning it is possible to minimise the effect of any single electrical fault on the overall precipitator performance and, by isolating an electrical sub-section, external repairs may be possible with the rest of the precipitator still in operation.

When siting the transformer/rectifier set it is important to minimise the extent of HT cabling which is expensive and sometimes troublesome. It may be practical to site the electrical sets on the roof of the precipitator itself and connect the HT bushing directly to the electrical discharge system with a solid metallic bus-bar arrangement, thus avoiding HT cabling entirely. Roof-mounted equipment may invite less maintenance inspection though and certainly requires a heavier support structure.

77

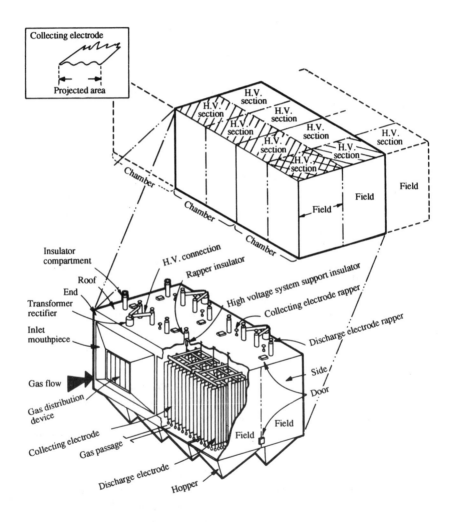

Figure 10.4   Layout of an electrostatic precipitator.

10.3   PRECIPITATOR PERFORMANCE

The behaviour of an electrostatic precipitator generally follows the formula proposed by Deutsch. There are many modified versions[52] of the Deutsch equation in current use, but the following basic exponential relationship between efficiency and design parameters is generally accepted:

$$1 - E = \exp\left[-\frac{WA}{V}\right]$$

where

E   is the efficiency of dust collection expressed as a fraction,
A   is the total surface area of the collector electrodes,
V   is the volumetric flowrate of the gas, and
W   is the effective migration velocity.

In its original concept the effective migration velocity was a measure of the speed at which charged dust particles migrated from the discharge wire to the collector surface. However, the many non-ideal factors which affect precipitation performance in practice are generally included into the effective migration velocity.

Rearranging the Deutsch equation illustrates the factors that affect the size of a precipitator:

$$A = \frac{V}{W} \times \ln\left[\frac{1}{1-E}\right]$$

The collector surface area, A, is clearly related to the size and cost of a precipitator and is dependent on the required collection efficiency.

In designing a precipitator, the gas volume and required efficiency are usually defined by the customer. The precipitator manufacturer selects a migration velocity based on his experience in similar applications, the chemical and physical analyses of the dust and, possibly, pilot plant data. It is unreliable to calculate the effective migration velocity, W, theoretically and it is invariably chosen from experience.

The value of the effective migration velocity is a direct measure of the precipitability of a particular dust and most typically varies from 4 cm/s to 15 cm/s depending on the process, particularly the physical properties of the dust.

To precipitate a dust electrically, the particles should have an electrical conductance within a certain specified range. In practice, it is more usual to define this property as the 'resistivity' of the dust. Resistivity is the inverse of conductivity and has units of ohm cm.

At low levels of resistivity, particles reaching the collector electrode lose their charge easily and dust re-entrainment can then occur. When the dust has too high a resistivity an insulating layer is formed on the electrodes which prevents normal corona discharge; this leads to electrical instability and lower collection efficiency. Most dusts do have a resistivity value within an acceptable range, although because resistivity is temperature dependent it is sometimes an advantage to cool the gas to an optimum level.

The injection of water for cooling hot gas results in an increase in the moisture content of the gas, and in itself can cause a decrease in the resistivity of the precipitated dust and consequently an improvement in precipitator performance. The injection of water primarily to reduce resistivity is called 'conditioning'.

It is also possible to 'condition' dusts which are difficult to precipitate by injection of small quantities of certain chemicals (usually SO_3 or NH_3) which modify the electrical conductivity of dusts. However, the effective injection of these chemicals can be difficult and expensive and is almost invariably limited to power station flue gas applications.

GAS DISTRIBUTION

To achieve the best performance from a precipitator it is essential that the gas flow through the unit is uniform and that no gas bypasses the electrical field. Uniform flow must be achieved at the inlet to the precipitator by correct design of inlet ducting and the use of flow distribution devices within the inlet mouthpiece. The flow distributors must be designed to spread the high velocity inlet gas over the full cross section of the precipitator itself and must achieve this with a minimal pressure drop. The flow distribution is generally achieved by a combination of splitters, baffles and/or perforated sheets.

Gas bypassing over the top of the electrical field, or underneath through the hopper area, is minimised by strategically positioned baffles.

To ensure correct flow distribution through precipitators most reputable manufacturers insist on a model study, usually on a 1/16 to 1/8 scale. Any maldistribution of flow can then be detected and rectified on the model and these modifications incorporated into the full-size plant design.

10.4   SAFETY REQUIREMENTS

EXPLOSION RELIEF

Relief panels or doors are provided in precipitator casings where explosions or pressure surges are possible. These are usually sited in the roof, rather than on the side of the casing where the internal arrangement of collector plates would severely limit their effectiveness. Casual access in the vicinity of the pressure relief panels must be prevented. If the relief area takes the form of hinged doors or flaps provision should be made to ensure that they re-seal following opening to eliminate any air leakage. For a typical precipitator, the explosion relief area required would be of the order of 5 m^2 per 100 m^3 of chamber volume. Detailed methods of calculation for specific applications are given in References 9 and 10.

TOXIC GASES OR DUST

Dust handling systems operating on toxic dusts must be completely sealed. Precipitators operating under pressure in cleaning toxic gases (such as blast furnace gas) should be stringently tested to ensure that the pressurised casing does not leak. If the precipitator can be isolated by gas-tight valves this can be readily quantified by monitoring the loss of pressure as a function of time. Assuming all joints have been fully proven, the first areas in which to look for leaks are around explosion and inspection doors, shaft seals and through dust valves.

HIGH VOLTAGE

All precipitators are fitted with a safety locking system to ensure that the rectifiers are switched off and the plant earthed before any access doors can be opened.

An additional precaution is the provision of an insulated, flexible earth clamp to enable any residual charge in the precipitator field to be earthed via the casing prior to entry.

10.5   MAINTENANCE REQUIREMENTS

Dust collectors do not generally contribute towards the output of the process on which they operate and consequently their maintenance is frequently given a low priority and often neglected. This is particularly undesirable for precipitators because a serious reduction in performance can result and, if faults are allowed to progress unchecked, repairs can be expensive.

Many processes operate 24 hours per day and are shut down as infrequently as possible for maintenance. To take advantage of these infrequent, short duration shutdowns it is essential that adequate maintenance access is provided to all key components. The main access doors should be quick-release, although safety-locked. Within the casing, walkways should be provided for inspection, maintenance and adjustment of the electrode system and the rapping assemblies. Access to the top plate of the precipitator must be provided to allow for convenient inspection and cleaning of the discharge electrode support insulators. For wet precipitators, access must be provided to the spray nozzle assemblies and the water distribution headers.

Facilities should be provided for inspection and minor adjustments of the dust removal drive mechanism from outside the precipitator, with the plant operating or with the drive isolated.

10.6   POSSIBLE PROBLEM AREAS

The various operational problems which can occur with electrostatic precipitators are described below. A prospective purchaser should satisfy himself that all proposals he receives make adequate provision to avoid these problems.

(i)   A frequent problem is over filling of hoppers, caused by:

• undersizing of hoppers, bearing in mind that the greatest dust fall out occurs in first field;

• inadequacy of conveyors or dust handling plant;

• shutdown of dust handling plant;

• use of hoppers as storage.

When hoppers fill up to the bottom level of the electrodes the high tension supply shorts out and makes the field inoperative. Also, if rapping continues, structural damage can occur to the electrodes which are no longer suspended freely.

(ii)   Discharge electrode breakage (more common with weighted wire type electrodes) which results in 'shorting-out' of the electrical sub-section until repaired.

(iii)   Tracking along insulators caused by dust or moisture build-up.

(iv)   Insulator breakage.

(v)   Poor gas distribution, gas bypassing and leakage through hoppers.

(vi)   Corrosion due to condensation caused by poor lagging or the inleakage of cold air.

(vii) Corrosion due to the gas temperature falling below dewpoint.

(viii) Dust build-up in ducts, inlet and outlet transitions.

(ix) Failure of rapping drives.

11. FABRIC FILTERS

Fabric filters are a widely used form of high-efficiency collector for both dust and fumes. Even on sub-micron dusts, fabric filters can give collection efficiencies of above 99%. They have the ability to handle wide variations in gas flowrate, dust concentration and particle size. This is particularly beneficial when the actual dust concentration and particle size analysis are not known accurately at the design stage. The development of synthetic filter cloths has enabled many different types of fabric filters to be used in a wide range of applications.

11.1 BASIC PRINCIPLES

The basic principle of fabric filtration is to select a fabric membrane which is permeable to gas but which will retain the dust. Initially, dust is deposited both on the surface fibres and within the depth of the fabric, but as the surface layer builds up it itself becomes the dominating filter medium. As the dust cake thickens, the resistance to gas flow increases. Periodic cleaning of the filter media is therefore necessary to control the gas pressure drop over the filter. The most common cleaning methods include reverse air flow, mechanical shaking, vibration and compressed air pulsing. Often a combination of these methods is employed. The normal cleaning mechanisms do not result in the fabric returning to its 'as new' condition. It would actually be undesirable to overclean the fabric because the particles deposited within the depth of the cloth help to reduce the pore size between the fibres, thus enabling high efficiencies to be achieved on sub-micron fumes.

Fabric filters are designed on the basis of an anticipated filtration velocity which is defined as the maximum acceptable gas velocity flowing through a unit area of fabric (expressed in m/s). Filtration velocities generally lie in the range 0.01 to 0.06 m/s according to the application, the filter type and the cloth. The vast majority of filters operate over a controlled cycle of filtration and regeneration such that the pressure drop will go no higher than 150 mm water gauge. Where the dust-initiating process operates intermittently, it is often

84

both convenient and cost effective to clean the fabric during the process down-time. Intermittent operation may also allow higher filter velocities to be used but care is required in sizing the dust handling plant and also to ensure that subsequent, more continuous, processing does not overload the filter.

The filter media are usually in the form of tubular bags, but other arrangements such as flat pockets are also widely used. Bags are typically 0.1 to 3.0 m in diameter and 2 to 10 m long. Gas flow can be in either direction so that, according to the filter design, the dust cake may build up on either the inside or the outside of the bag. With inward gas flow, the bags are prevented from collapsing by a wire or plastic internal cage.

11.2 PRESSURE DROP ACROSS THE FILTER

Because it is a highly efficient type of collector, it is unusual for a fabric filter not to meet the requisite particle emission limit. The reason for failure can usually be traced to a pinhole or tear in the fabric, lack of integrity in a seal or a joint, overcleaning of the fabric, etc. Where fabric filters do fail to achieve the desired performance it is usually the result of an excessive pressure drop across the fabric causing the gas flowrate to fall below the design value. This in turn may cause a pronounced deterioration in the performance of main process plant (such as a dryer) or a serious low-level emission of dust and fume from around the dust source. It is important, therefore, to understand what factors affect pressure drop and how pressure drop is likely to change with time.

The pressure drop (ΔP) across the filter can be expressed as:

$$\Delta P = \Delta P_f + \Delta P_d$$

ΔP_f is the pressure drop across the 'used' fabric (ie with the fabric containing a residual or interstitial deposit of dust immediately after cleaning) and ΔP_d is the pressure drop across the dust deposit which increases with time, raising ΔP to the point where the fabric requires to be cleaned. The above equation can also be expressed as:

$$\Delta P = K_1 v + K_2 v\, w$$

or $\Delta P = K_1 v + K_2 v\, (cvt\, \eta)$

where v is the filtration velocity or 'air-to-cloth ratio'
w is the mass of dust deposit per unit area or 'areal density'
c is the concentration of dust in the gas stream

85

η is the fractional collection efficiency (usually assumed to be unity) and t is the time since the fabric was last cleaned.

The above equation assumes c, v and η are constant with time. K_1 is characteristic of the cloth/dust system and the method used to clean the fabric and K_2 is characteristic of the dust deposit. Both factors are also dependent upon the temperature and humidity of the gas. K_1 and K_2 cannot be estimated from first principles with any reasonable degree of accuracy and have to be obtained experimentally. It follows from the above equation that a plot of ΔP against t will be linear; the slope of the plot is proportional to K_2 whilst extrapolation to $t = 0$ gives an intercept from which K_1 may be calculated. The equation can be applied to a multi-compartment baghouse[60] (see Section 11.5), but because of the problem of dust redeposition it should not be used with a filter employing on-line pulse-jet cleaning (see Section 11.4).

11.3   FABRIC SELECTION

The correct choice of fabric is most critical in designing a new fabric filter and the selection is usually made by the equipment supplier, in conjunction with his fabric suppliers. A general guide to the selection of the correct type of filter cloth is shown in Table 11.1.

Fabric selection must take into account the composition of the gases, the nature and particle size of the dust, the method of cleaning to be employed, the required capture efficiency and economics. The gas temperature must also be considered, together with the method of gas cooling, if any, and the resultant water vapour and acid dewpoint. Methods of gas cooling are discussed in Chapter 7 and in Reference 28.

Characteristics of the fabric to be considered include chemical resistance of the fibre, fibre form and type of yarn, fabric weave, fabric finish, abrasion and flex resistance, strength, collecting efficiency, cloth finishes and cloth permeability.

Both natural and synthetic fibres can be produced in the form of staple (many short lengths of fibre, typically 1 to 10 cm long), monofilament (a single length of long fibre >>10 cm) or as multifilament (many continuous, parallel, long fibres). Fabrics are made from yarn, which is a generic term for a continuous strand of textile fibres or filaments, with or without twist. Yarns formed of short-staple fibres are called spun yarns; these require sufficient twist to hold the fibres in place.

TABLE 11.1
Typical filter fabrics

Fabric	Maximum recommended operating temperature	Acid resistance	Alkali resistance	Abrasion resistance	Typical applications
Cotton	82°C	Poor	Very good	Very good	Ambient temp: non-acid or unusually abrasive dust
Wool	93°C	Good	Poor	Good	Smelters
Nylon	93°C	Fair	Excellent	Excellent	Low temp: highly alkali or highly abrasive dust
Polypropylene	93°C	Excellent	Excellent	Excellent	Low temp chemical application
Acrylic	127°C	Good for mineral acids	Good for weak alkalis	Good	Steel furnaces, chemical smelters
Polyester	149°C	Fair	Good	Good	Steel furnaces, chemical application
PTFE	260°C emits toxic gas 232°C	Inert except for fluorine	Inert most alkalis	Fair	Chemicals
Aramid	204°C	Fair/good	Good	Good	Chemical, non-ferrous chemicals
Glass fibre	260°C	Fair/good	Fair/good	Fair/good	Carbon black, cement, steel furnace, non-ferrous chemicals

Most of the cloths used in early forms of fabric filter were woven. Such fabrics are constructed from two sets of yarns positioned at right angles to one another. Depending on how they are then interlaced, the resultant weave is referred to as plain, satin, twill, etc. The more common types of fabric construction are shown in Figure 11.1 overleaf.

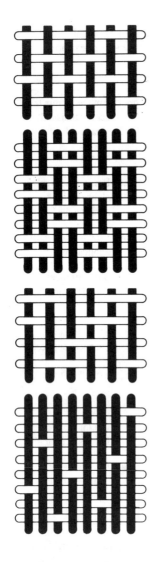

PLAIN

The threads interlace in alternate order, each warp or weft thread passing over one pick or end and under the next. This produces a firm fabric with maximum resistance to distortion and abrasion

2/2 MATT

In both directions there are two or more threads woven as one thread eg a 2/2 matt is double end, double pick, plain fabric. It is slightly stronger than a plain cloth and yet still retains firmness.

2/2 TWILL

The warp or weft threads float over the crossing threads for two or more picks or ends respectively. The points of intersection move outward and upward on successive picks, to form diagonal lines. Common twills are 2/2 and 3/1.

8 END SATIN

The surface of the cloth consists almost entirely of warp or weft floats, since in the repeat of a weave each thread passes over all but one thread in the opposite direction. Satin weave imparts good drape qualities and resistance to stretch, but has little firmness.

Figure 11.1   Common forms of fabric construction.

Most fabrics used in filtration are first woven and then given further treatment. For applications involving fine fumes, a raised nap finish produces a random fibrous surface which prevents small particles irreversibly penetrating deeply into the pores of the cloth which cause cloth 'blinding'.

Another popular class of filter fabric is referred to as needle felt. This is a random distribution of fine, staple fibres mechanically interlocked by the textile needle-punching process. A relatively lightweight (75 to 150 g/m^2) and usually open mesh woven cloth, or scrim, is embedded in the felt to improve the tensile characteristics, although this scrim does not usually play any part in the filtration mechanism of the felt. Needle felts are normally 300 to 700 g/m^2 weight and 1.5 to 3.0 mm thickness. The fine, staple fibres from which they are made are normally 10 to 20 microns in diameter and circular in cross section. They are mechanically interlocked to produce needle felt densities of about 0.20 to 0.30 g/cm^3 which in turn gives mean pore sizes of about 15 to 35 microns. The pore size varies considerably which partly explains why needle felts can achieve very high collection efficiencies even on sub-micron particles.

The trend in filter fabrics is away from woven fabrics and towards needle felts, mainly because needle felts allow for increased filter velocities and usually give higher collection efficiencies. Table 11.2 on pages 90 and 91 shows how needle felts will filter at about twice the velocities possible with woven cloths in the same dust application.

Needle felts can be made electrically conductive (antistatic) by the inclusion of metal or carbon-coated fibres which reduces the explosion hazard in fabric filters. Needle felts can also be treated with a range of chemicals to improve their water and oil repellancy, acid resistance, collection efficiency and resistance to blinding. Needle felts made from glass, stainless steel and ceramic fibres are being developed to give higher temperature resistance. Special surface finishes are available to improve cake discharge characteristics and these sometimes also improve their resistance to blinding. Another special surface is achieved by bonding a thin layer of micro-porous PTFE on to the surface of a standard polyester or aramid felt.

Particular types of cloth are often associated with particular types of filter. Woven-type fabrics are normally used in shaker and low-pressure, reverse-flow type filters and generally operate at filtering velocities between 0.01 and 0.03 m/s. In pulse-jet cleaning filters, needle felts are preferred, operating at filtering velocities of 0.02 to 0.06 m/s.

Table 11.2 gives some guidelines on filtration velocities for various

TABLE 11.2
Typical filtration velocities on dust applications for fabric filters

Dust	Recommended average filtration velocities	
	woven media m/s	felted media m/s
Aluminia	0.0152	0.0305
Aluminium oxide	0.0165	0.0381
Ammonium chloride	0.0102	0.0229
Aminal feeds	0.0203	0.0508
Asbestos coarse	0.0267	0.0483
Asbestos fines	0.0165	0.0406
Bakelite	0.0178	0.0381
Bauxite	0.0203	0.0406
Boric acid	0.0203	0.0406
Brick grog	0.0229	0.0483
Carbon black	0.0127	0.0254
Cast iron dust	0.0203	0.0457
Cement	0.0165	0.0381
Ceramics	0.0165	0.0381
Charcoal	0.0203	0.0508
Clay	0.0165	0.0356
Coal	0.0165	0.0356
Coffee	0.0178	0.0406
Coke	0.0178	0.0279
Copper	0.0165	0.0381
Copper ore	0.0178	0.0406
Cork	0.0229	0.0508
Corn starch	0.0127	0.0356
Cosmetics	0.0127	0.0356
Detergents	0.0152	0.0254
Diatomaceous earths	0.0178	0.0305
Epoxy resin	0.0127	0.0254
Felspar	0.0152	0.0381
Fertiliser	0.0152	0.0381
Flint	0.0152	0.0356
Flour	0.0203	0.0406
Fly ash handling	0.0152	0.0305
Fullers earth	0.0152	0.0381
Glass	0.0152	0.0330
Grain dust	0.0203	0.0559
Granite	0.0203	0.0406
Graphite	0.0152	0.0305
Gypsum	0.0152	0.0356
Iron ore	0.0165	0.0381

TABLE 11.2 (continued)
Typical filtration velocities on dust applications for fabric filters

Dust	Recommended average filtration velocities	
	woven media m/s	felted media m/s
Kaolin	0.0165	0.0279
Lamp black	0.0127	0.0305
Lead oxide	0.0127	0.0305
Leather	0.0203	0.0381
Lime hydrated	0.0152	0.0356
Limestone	0.0152	0.0330
Magnesia	0.0203	0.0432
Marble	0.0203	0.0432
Metal powders	0.0152	0.0254
Mica	0.0165	0.0330
Milk powder	0.0165	0.0330
Oxides (metallic)	0.0152	0.0330
Paint pigment	0.0127	0.0254
Paper	0.0254	0.0508
Perlite	0.0165	0.0356
Pharmaceuticals	0.0165	0.0406
Plastics	0.0165	0.0356
Pottery (clay)	0.0114	0.0356
Pumice	0.0165	0.0356
Quartz	0.0165	0.0356
Refractory dust	0.0165	0.0356
Rubber chemicals	0.0127	0.0254
Salt	0.0165	0.0356
Sand	0.0229	0.0457
Sander dust	0.0305	0.0610
Shotblast dust	0.0178	0.0406
Silica flour	0.0165	0.0330
Slag	0.0165	0.0356
Soap dust	0.0127	0.0254
Soda ash	0.0165	0.0356
Soya bean flour	0.0165	0.0356
Sugar	0.0127	0.0381
Sulphur	0.0102	0.0305
Talc powder	0.0127	0.0305
Titanium dioxide	—	0.0178
Tobacco	0.0254	0.0508
Trisodium phosphate	0.0165	0.0406
Whiting	0.0127	0.0330
Zinc carbonate	0.0127	0.0330

applications. Whilst some felt cloths can be used in the shaker and low-energy, reverse-flow cleaning filters, care has to be taken because of the greater tendency for the cloth to blind. Due to the greater cleaning force generated in the pulse-type filter, felts are invariably used with this type of plant.

Clean filter fabrics may be compared by measuring their permeability. This is defined as the volumetric flowrate of air passing through unit area of the cloth at a given pressure drop (usually 12 mm water gauge). In practice, permeability values should be used with caution in fabric selection because they do not give an accurate reflection of the fabric performance when coated with dust.

11.4   TYPES OF FABRIC FILTER

Fabric filters are normally classified according to the method by which the filter media are cleaned. Regular dust removal from the fabric is important in order to maintain effective extraction efficiency, but it also influences the operating life of the fabric.

There are many different types of cleaning, but most can generally be classified as one or a combination of the following:

• reverse flow;

• mechanical shaking;

• pulse-jet.

REVERSE-FLOW FILTERS

Where a smooth-finish fabric can be used (eg terylene or fibreglass in high temperature applications) automatic cleaning can be carried out by a simple reverse flow of air provided by an auxiliary fan as shown in Figure 11.2.

On the filter shown, the dust-laden gas enters the hopper inlet and passes upwards through the inside of the filter tubes. At regular intervals, the clean gas outlet damper on one compartment is closed, stopping the gas flow through that compartment; at the same time, the reverse-flow system damper is opened and the reverse air flows through the compartment. This causes the filter tubes to collapse, discharging the dust deposit from the inside of the tubes into the hopper below. After a short time, the gas outlet damper opens and the reverse-flow damper closes bringing the compartment back on line. The cycle is repeated for each compartment in turn. Further details on the operation of multi-compartment filters are given in Section 11.5.

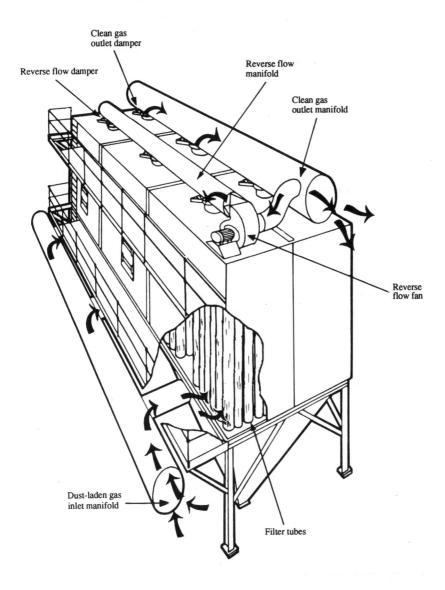

Figure 11.2 Typical reverse flow filter.

MECHANICAL-SHAKER/FILTERS

Mechanical shaking and rapping involve the use of a mechanical device such as a rocker arm and lever assembly to produce an oscillatory motion at the top of the filter tubes. This imparts a rippling wave action down the fabric causing the dust deposit to fall off into the hopper below. The motion applied at the top may be essentially horizontal, vertical, or both (covering a 90° arc). Vertical motion is sometimes accompanied by rapping. The fabric is supported at the top and the dust deposit forms on the inside of the tubes (or pockets). For maximum effectiveness, the air or gas flow must be stopped when the fabric is to be shaken and hence, for continuous operation, a multi-compartment filter is used (further details given in Section 11.5).

Mechanical shaking is a more effective cleaning mechanism than reverse-flow, generally resulting in lower pressure drops and/or higher filtration velocities across the filter. The disadvantages in mechanical shaking are the mechanical complexity of the shaking gear and the heavy wear at the fixed base of the tubes due to continual flexing, particle abrasion, etc.

PULSE-JET FILTERS

The pulse-jet filter consists essentially of a series of filter elements supported at the top by a cell plate and enclosed in a dust-tight, fabricated metal housing as shown in Figure 11.3. Dust is deposited on the outside of the filter elements and is removed at regular intervals by a powerful jet of compressed air directed into a venturi tube which is located in the top of each filter element. The 'venturi effect' (based on the same principle as the jet pump) induces more air into the tube from the clean air plenum, thus magnifying the intensity of the cleaning action. The air pulse momentarily stops filtration, expands the filter element rapidly and causes gas flow in the reverse direction from the clean side of the fabric to the dirty side. The rapid movement and reverse flow of gas dislodges the dust deposit from the fabric and the agglomerated dust particles fall into the hopper below. Each row (or pair of rows) of filter elements is cleaned in turn while the gas or air continues to flow through the remaining elements. The pulsing is controlled by diaphragm valves on the compressed air manifolds. These valves are operated by solenoid pilot valves which are in turn actuated by a solid-state electronic timer. The timer has provision for adjusting the frequency and duration of the pulse. Typically, the duration of the pulse would be set somewhere between 50 and 150 milliseconds and the pulsing interval at 10 seconds, but these might be adjusted during commissioning of the plant to give optimum performance.

A wide variety of filter designs is available. The filter elements may take the form of either tubes or pockets. There is also a choice of gas inlet position. The dust-laden gas may enter the bottom of the filter below the filter media (as shown in Figure 11.3) in order to allow the coarser particles to settle

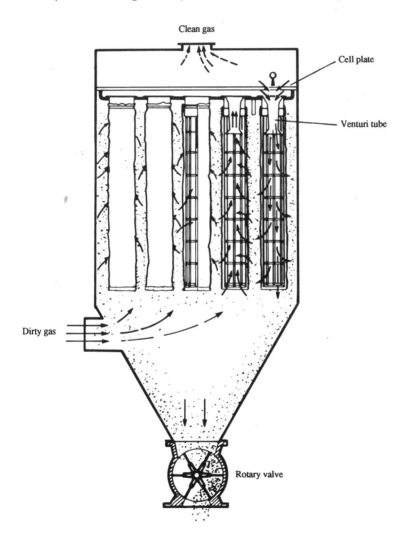

Figure 11.3 Typical pulse-jet filter.

95

immediately into the hopper. Alternatively, with finer dusts, the gas may enter the top of the filter just below the cell plate; the downward flow of gas generally results in more effective cleaning and hence in lower pressure drops. Access to the filter is provided either by a 'walk-in' chamber above the cell plate or by providing the clean air plenum with lift-off doors on the top, thereby allowing inspection and replacement of tubes, support cages, etc entirely from the clean side.

The chief characteristics of pulse-jet filters are as follows:
(i)   the dust is collected on the outside of the filter medium;
(ii) needle-felt fabrics are generally used instead of woven cloth;
(iii) high filtration velocities are normally used;
(iv) compressed air is used for cleaning, thereby avoiding the use of moving parts;
(v)   cleaning usually takes place with the filter elements on-line (although off-line cleaning is also used, particularly where dust re-entrainment is likely to be a problem).

11.5   FILTER ARRANGEMENTS

UNIT FILTERS

When a single or a number of small fabric filters are located adjacent to the source of the dust, this is referred to as a unit solution. Such installations give savings in ducting and offer a high degree of convenience. Unit filters are generally restricted to dust collecting applications where they are easy to install, operate, service and redeploy. Where no health hazard can arise, it is sometimes permissible to discharge the cleaned air back into the work area.

Dust disposal from a unit filter, particularly if several are involved throughout a plant, must be given careful consideration. The normal practice of bagging and emptying of bins and trays into a container may not be acceptable if it causes a secondary dust nuisance or a high labour cost. Regular inspection and emptying of containers is essential.

Unit collectors are available in a wide variety of arrangements and sizes. Figure 11.4 shows a typical unit dust collector.

Another development is to use a unit filter without a casing and hopper by inserting the pockets or tubes directly into the unit being vented, such as a silo or an enclosed conveyor system transfer point (see Figure 11.5, page 98). With unit filters, cleaning is generally by pulse-jet or mechanical shaking.

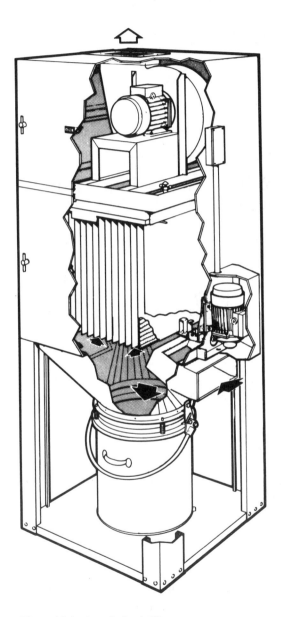

Figure 11.4   A typical unit filter.

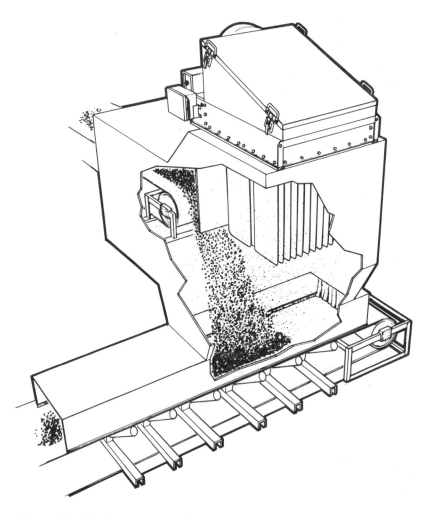

Figure 11.5   Unit filter incorporated into conveyor transfer point.

LARGE, SINGLE-COMPARTMENT FILTERS
For the larger and/or more difficult applications or where dust-laden gases are
being extracted from a number of different sources to a single dust collection
point (often referred to as a 'centralised plant solution') a large single-compartment filter is frequently employed. Pulse jet cleaning is the most commonly used
method of cleaning in such filters although other methods are sometimes used
such as reverse flow (using an external blow ring or a travelling carriage).

To save costly production space, it is common practice to house the filter, fan and dust handling system outside the main working area. Maintenance is then confined to one area and dust disposal can be automated so that it does not create a secondary dust nuisance.

MULTI-COMPARTMENT FILTERS

An alternative to the large single-compartment filter is the multi-compartment filter in which the filter elements are cleaned off-line. In this type of unit, a damper is incorporated in the outlet duct from each compartment; by closing the damper, the compartment can be isolated from the main gas stream for cleaning. Shaker-type, reverse-flow or pulse-jet cleaning can be used depending upon the application.

The operation of a multi-compartment filter is complex and requires further explanation. Figure 11.6 shows a typical arrangement for a five-compartment filter in which compartment no. 5 is off-line for cleaning. Figure 11.7 overleaf shows the change in gas flowrate through each compartment from the time when compartment no. 1 has just come back on stream. After a certain period, compartment no. 5 goes off stream as indicated in Figure 11.6. The cleaning sequence is 1 → 5 → 4 → 3, etc. The amount of dust deposit on the bags in the various compartments is different and depends on the time that has elapsed since the bags were last cleaned. Because there is the same pressure drop over each 'on-line' compartment, when a newly cleaned compartment comes on stream the flowrate through that compartment is very high initially but falls as the new dust layer is deposited, whilst the gas flowrates through the

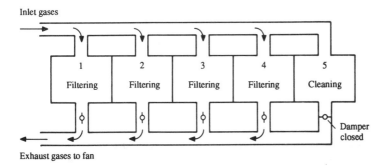

Figure 11.6   Typical arrangement of a multi-compartment baghouse.

99

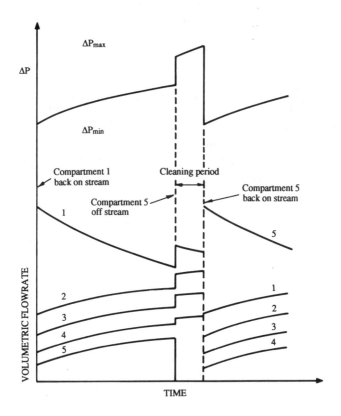

Figure 11.7   Variation of volumetric flowrate and pressure drop with time in a
multi-compartment baghouse during two consecutive cleaning cycles.

other compartments increase to compensate. When compartment no. 5 is
removed for cleaning, the gas flowrates increase abruptly through the others.
Figure 11.7 also shows the variation with time of the pressure drop across the
filter. The pressure drop increases abruptly when one compartment is removed
for cleaning and falls abruptly when the newly cleaned compartment comes back
on stream. The greater the number of compartments in the baghouse, the less is
the increase in pressure drop when a compartment goes off-line.

It is good practice to install a differential pressure switch across the
filter which actuates the cleaning cycle on reaching a preset pressure drop. In
this way, the bags are cleaned only when required, thus preventing overcleaning,

prolonging bag life and minimising compressed air usage, heating and other costs.

OPEN-TYPE FILTERS

An attractive design for larger volume installations, when it is permissible to discharge to atmosphere out of the roof of the filter, is to house the filter tubes in a simple weather-clad enclosure or building, forming what is known as an open-type pressure filter. The filter tubes are arranged in sections to allow isolation for cleaning, each section having its own dust collection hopper. The air or gas is blown under pressure into the hoppers, through the filter tubes and out to atmosphere via a top-ridge ventilator in the building. The dust is deposited on the inside of the filter tubes which are cleaned by reverse flow or by mechanical shaking. This type of filter is becoming less widely used.

11.6   ADVANTAGES AND DISADVANTAGES OF FABRIC FILTERS

There are many advantages and disadvantages in using fabric filters. These are summarised below.

ADVANTAGES

- Very high collection efficiencies.
- Dry, low visibility plume.
- No effluent treatment required.
- Power costs moderate.
- Air can often be recirculated.
- Corrosion not usually a serious problem.
- Collection of product in dry condition.
- Low capital cost on simpler applications.

DISADVANTAGES

- Gas cooling often required, which is difficult and expensive.
- High capital cost on 'difficult' processes.
- High maintenance costs on 'difficult' processes.
- Fabric conditioning sometimes required.
- Expensive fabrics sometimes required.
- Secondary dust problems.
- Dewpoint problems leading to blinding of fabric.

101

- Fire and explosion hazard.
- Cleaning air (reverse flow) sometimes requires to be heated.
- On some applications, hot air recirculation system required to protect filter against condensation during downtime.
- Some dusts very difficult to dislodge from fabric, causing the filter pressure drop to exceed the design value.
- Bag replacement required periodically.

11.7   MAINTENANCE REQUIREMENTS

EASE OF INSPECTION (INTERNAL OR EXTERNAL)

Inspection of filters is an essential prerequisite to effective maintenance. It should be remembered that such tasks are often performed by unskilled operators. It is essential, therefore, that inspection of any filter should be simple, straightforward, safe and present no danger to health.

ACCESS

Access should be easy and safe for all personnel authorised to enter the filter. Doors should be of robust design with good seals and practical fastening devices, interlocked with related equipment. All potentially hazardous operations should have suitable warning notices and instructions as required to personnel concerned. Safety steps should be provided rather than safety ladders. Fastening attachments for safety harnesses should be provided where required.

FILTER ELEMENT FAILURES

As filter elements generally give a good operating life (one to four years is typical depending on application and type of plant) there is a tendency for maintenance to lapse. This can be an expensive error as once one or two elements fail the escaping dust can contaminate adjacent elements and failures soon spread. Regular inspection for signs of wear and correct tensioning and location of the filter elements should be made.

COMPRESSED AIR

Compressed air supply for use on pneumatic actuators on dampers and also for the high energy cleaning units should be clean, relatively water free and available at a pressure of 4 to 7 bar. Compressed air requirements vary for different types of filter and this may affect both the capital and operating cost of the plant.

FAN SETS

Fabric filters generally provide a high dust collection efficiency and it is normally possible to install a high efficiency fan on the clean side of the filter. This saves power and reduces operating and maintenance costs and may provide lower noise levels. Further information on fan selection is given in Chapter 14.

12. DEEP BED FILTERS

While the most common form of gas filtration utilises fabric filter media, there are several other types of filter which have importance in specific applications. One of these, paper filters, is only infrequently used for industrial pollution control as it is more suited for small gas flowrates containing very low dust concentrations (mg/m^3). Of greater industrial importance are fibrous mat filters and aggregate bed filters.

12.1   TYPES OF FIBROUS FILTERS

Fibrous filters are used to clean gases in which the dust concentration is low. They consist basically of fibrous mats through which the dusty gas is passed. The particles are deposited on individual fibres where they are retained within the depth of the mat. When the deposited particulates within the filter restrict the gas flow to an unacceptable level the mat is normally replaced, although sometimes it may be washed.

Typical applications include dust removal in air conditioning systems, in engine intakes, in face masks and in other situations where an extremely high efficiency of particulate removal is necessary, such as in the nuclear industry. Such filters can be broadly classified as either coarse filters, which are primarily used to remove larger particulates, or high-efficiency filters, which are used in the efficient removal of fine aerosols. Typical characteristics of these two classes are shown in Table 12.1.

A wide range of fibrous media is available to suit diverse applications and many sub-divisions of the above classification could be included. 'Roughing filters' which may be used as primary collectors upstream of a high-efficiency filter normally have a very low pressure drop (mm W.G.) and retain very few particles below 3 m diameter. High-efficiency units which include the HEPA (High Efficiency Particulate Air) and UHEPA (Ultra High Efficiency Particulate Air) types contain compressed media arranged in concertina configurations to increase the face area for a given unit volume. The efficiency can be extremely high and the pressure drop may reach 50 to 70 mm W.G.. Both types of unit are

TABLE 12.1
Classification of fibrous filter systems

	Efficiency of collection of 0.3 μm particles	Fibre diameter (μm)
Coarse filters	50%	300–800
High efficiency filters	95–99.997%	1–20

	Packing density (fibre volume per unit mat volume)	Face velocity (m/s)
Coarse filters	0.03	0.5–3.0
High efficiency filters	0.10	0.03–0.10

normally assembled in conveniently sized panels such as 0.5 × 0.5 m (see Figure 12.1) or in cannisters which can be installed in utility gas lines.

In an alternative arrangement, the filtering media can be in the form of a continuous roll. Coarse media may be washed in an oil bath at the base of the unit and recycled. Finer dry media may be replaced using the roll principle as

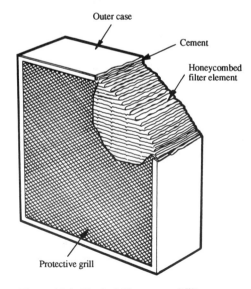

Outer case

Cement

Honeycombed filter element

Protective grill

Figure 12.1   Typical fibrous panel filter.

105

the pressure drop in a system reaches an unacceptable level. This may be particularly useful where hazardous gases are treated.

12.2   FIBROUS FILTER PERFORMANCE

By considering the total area of fibre exposed to the gas as it passes through the media, it can be shown that the particle penetration falls exponentially with increase in the surface area of fibre exposed for collection. The overall collection efficiency is thus increased as the depth and packing density are increased and the fibre size reduced.

Several different mechanisms are responsible for collection. With small particles, approximately less than 0.5 µm diameter, the predominant mechanism is Brownian diffusion which is most effective with low air velocities which permit adequate residence time in the proximity of the fibre surface. Small fibres are most conducive to high efficiency in this regime. With larger particles, inertial interception is dominant and a high Stokes' Number (see Chapter 8) leads to a high efficiency. In both these regimes electrostatic attraction may significantly enhance efficiency and some media are now manufactured which have strong electrostatic properties; their main application is in face masks. Particle-fibre adhesion is by Van der Waals molecular forces, although with dry filters electrostatic attraction may be significant. Where large particles (5 µm) impinge on the fibre surface they may bounce off causing a reduction in collection efficiency. However, in industrial applications, this is often obviated by coating the filter with an adhesion-assisting fluid.

With high-efficiency filters designed to collect particles in the respirable size range and below, the use of fine fibres is beneficial. The uniformity of fibre packing in the mat is important and the presence of a few 'pinholes' can significantly increase particle penetration. With filters designed to remove coarser particles, larger fibres are used for their mechanical strength. The adhesion efficiency is also increased because the particle fibre impact velocity is reduced.

The pressure drop across a filter is proportional to the surface area of fibres exposed and is almost proportional to the face velocity. With high-efficiency filters, the efficiency normally increases with the dust load within the filter, as the dust itself acts as an efficient filter. The pressure drop for a given face velocity always increases with dust loading; this is normally the limiting factor in determining the useful life of a filter. Careful maintenance is needed to establish the point at which filter replacement is necessary.

12.3   FIBROUS FILTER TESTING

A range of standard test procedures is available and fibrous filters are normally classified in accordance with their performance against one or more of these. The more important test methods are listed together with their range of applicability in Table 12.2 on pages 108 and 109. The most important of these for high-efficiency filters are the sodium flame test in which fine salt aerosol is used (UK), the DOP test (USA) and the oil mist test (Germany). With coarse filters, standard test dusts are used to measure overall collection efficiency and the dust retention capacity of the filter. It is essential that the performance against the appropriate tests should be quoted in specifying the requirements of a filter system.

12.4   AGGREGATE BED FILTERS

Although the number of aggregate bed filters currently in use is small, they would appear to have potential for those applications which are less well suited to the fabric filter. Thus, the aggregate filter has been applied to high-temperature duties and to applications involving chemically reactive dusts. Like the fabric filter, however, care must be taken to avoid operation close to dewpoint temperatures which would lead to the formation of tenacious dust deposits. This is particularly important on start-up because aggregate filters have a high thermal capacity which can cause a large temperature reduction of the gas during initial operation.

The most common form of aggregate filter utilises a bed of gravel or sand, although many other materials have been tried including coal, coke and soil. Sometimes the bed is fixed and remains in use until the dust deposits cause an unacceptable resistance to gas flow. The bed is then discharged and replaced by fresh aggregate. In the more modern form of aggregate filter, however, different sections of the bed can be regenerated either mechanically or by a fluidising medium which passes through the bed in a counter-current flow.

Aggregate filters may comprise beds several metres in depth and, if fine sand is used, efficiencies over 99% can be achieved even on sub-micron particles. Sometimes the bed is graded in particle size so that the highest dust deposits are contained in the coarser regions of the bed. This reduces the tendency for the finer sand to become quickly blocked with dust.

TABLE 12.2
Methods of testing fibrous filter performance

Test method	Particle diameter (microns)	Size range	Suitability for *in situ* testing
Spot tests	Variable 0.1–1.0	Wide (atmospheric)	Poor
Radioactive	0.3	—	Good
DOP	0.3	Narrow	Good
Methylene blue	0.5	0.03–1.2	Poor
Sodium flare	0.6	0.03–1.5	Could be developed
Ultramarine	4.7	Wide	—
BS2831 dust no 2	7.5	Wide	Good
BS2831 dust no 3	22.5	Wide	Good
BS1701	50	Wide	Good

TABLE 12.2 (continued)
Methods of testing fibrous filter performance

British standard	Comments	Method of generation	Method of detection
—	Uses atmospheric dust	—	Stain
—	—	—	Scintillation counter
—	Popular in USA, high capital cost, sensitive, not suitable at very high temperature	Sinclair La Mer generator	Light scattering
BS2831	Popular, cheap, low sensitivity	Sinclair La Mer generator	Stain
BS2938	Popular British method, moderate capital cost, very sensitive, suitable up to 300°C	Collision generator	Flame photometer
—	—	Dry dispersion	Flame photometer
BS2831	Popular and cheap	Dry dispersion	Gravimetric
BS2831	Popular and cheap	Dry dispersion	Gravimetric
BS1701	Popular and cheap	Dry dispersion	Gravimetric

13. HANDLING AND DISPOSAL OF DUST AND SLURRIES

Depending on its application, the dust collector may capture dust at a rate from a few kilograms up to many tonnes per day. Whether the collector is a wet or dry type, it is normally good practice to discharge this material from the collector on a regular basis because large volumes of compacted dusts or slurries become particularly difficult to handle and in some instances may present a fire hazard.

13.1 CONTROLLED DISPOSAL

The problems of dust disposal become more difficult as plants tend to get larger and both industry, and its neighbours, become more environmentally conscious. Wherever possible, consideration should be given to returning the dry dust or dewatered sludge back into the process from which it derived. Alternatively, it may be feasible to find an outlet for the material for some other industrial use. In practice, however, the majority of dusts and sludges will be regarded as waste, but this does not mean that they can be forgotten. For most factories, on-site disposal space will be limited. In addition, extreme caution is required; uncontrolled tipping is dangerous and illegal, even on ground owned by the producer.

To obtain advice on the best means of disposal of any particular material, contact should be made with reputable waste management contractors or local authority or government services. They will need to know the type and quantity of dust or sludge involved, its water content and details of any known toxic or hazardous constituent. Even when the bulk of the waste is believed to be innocuous, laboratory analysis of the dust should be conducted to confirm that none of the minor components are harmful. Contractors will usually provide skips, which they will transport away on a regular basis, or they will send a tanker to empty a user's sludge tank.

It is often not possible to export the waste in the form in which it first emerges from the dust collector. Dry dusts may need to be bagged or wetted so that they are less prone to becoming redispersed in a wind. The discharge from a wet collector would probably represent too large a volume in its dilute form and may need to be dewatered prior to disposal.

In the following sections, general guidance is provided on the preparatory processes which may be required prior to the disposal of collected dust.

13.2   HOPPER DESIGN

Hoppers must be designed for free flow of dust and the minimum valley angle will be selected from experience and dust property tests. Hoppers are generally of the pyramid type, although trough types are sometimes used. When assessing the capacity of a hopper, the maximum dust loading condition must be accounted for, particularly in inlet hoppers where the greatest proportion of dust will fall out.

In general, the use of hoppers as storage capacity should be minimised as overfilling can cause serious operational problems in most collectors. The whole dust collection system must be carefully integrated with the cleaning cycle of the dust collector and sized for peak demand. Undersizing of hoppers is a major cause of problems.

Undersizing of the outlet valve at the base of the hopper should be avoided because this will not only increase the residence time of the dust in the hopper, but will also promote dust 'bridging'. A good quality valve is imperative to prevent inleakage of air into the plant. Trough hoppers usually incorporate a completely sealed conveyor in the base, which terminates in an outlet valve at one end. Again, this valve should be gastight.

The dust from a collector may be finer than that normally associated with the main process. This may cause it to be more cohesive and thus more difficult to handle than expected.

Mechanical vibrators or air fluidisation can be used to assist the flow of difficult dusts. Trace heating and lagging are provided on the hoppers if a temperature drop is likely to cause stickiness and blocking.

13.3   HANDLING AND CONDITIONING OF DRY DUSTS

Where only limited quantities of dust have to be disposed, the simplest and cheapest technique is to dump the dust direct from the collection hopper into a bag. If this cannot be done when the dust collection plant is running, then the hopper outlets or collection screw outlet should be fitted with a rotary discharge valve. It is also advisable to fit a twin bagging unit on the dust valve so that full bags can be removed without stopping and starting the output.

As an alternative to bagging, some dusts may be directly dumped into disposal containers, or lorries, if the material is first wetted. For this to be

111

effective, thorough mixing of the dust and water is essential. One way to achieve this is by the use of a twin-screw paddle mixer. Such equipment is best located close to the hopper outlet of the collector so that there is minimal risk of spillage or wind-blown loss prior to wetting. The consistency of the final mix may need to be varied to suit requirements by manual adjustment of the water sprays in the top of the mixer trough.

Where there are a number of dust outlets, it is sensible to have a collection system (mechanical or pneumatic) to transfer the dust into one storage hopper, and then fit one wet mixer unit on to the hopper outlet. Items such as bin level indicators with alarms, activated panels in the hopper to assist dust flow and dust feeders to control the rate of flow into the mixer are all improvements being used to assist productivity.

On large plant, where greater transfer distances are involved, pneumatic conveying systems are sometimes employed. The collected material should be removed at regular intervals from the dust collector hoppers using air slide conveyors and/or rotary valves. The material is then collected in a vacuum conveying system and deposited into a central storage silo. The maintenance on pneumatic conveying systems is directly related to the designer's skill in minimising erosion, particularly at bends. The silo exhaust may need to be continuously vented through suitable filters. Over shorter distances belt or drag link conveyors may be employed but it is important that both the conveyor and the transfer points should be totally enclosed to prevent wind losses.

Where the collected material has to be re-handled with the minimum risk of secondary dust nuisance, it may be worthwhile adding a binder to the dust and then pelletising or extruding the mix to form tablets, ovoids or logs.

13.4   THE HANDLING AND THICKENING OF WET EFFLUENTS

If fed with relatively clean water, a typical wet deduster is likely to discharge dirty liquor of concentration between 0.5% and 5.0% w/w solids. For all but the smallest systems, it would be wasteful on water and environmentally unacceptable to continuously reject such an effluent into drains or waterways. There may also be an unacceptably high cost in either transporting or discharging to sewers large volumes of weak effluent. Such action would also preclude the possibility of recovering any constituent of value from the effluent. A degree of concentration can be achieved by total or partial recycling of the scrubber wash liquor, but simple calculation quickly shows that it would not take many cycles to produce a slurry that was difficult to pump and prone to pipeline blockage. Hence it is necessary to provide a means of separating some of the wash liquor from the collected dust.

The most common means of separating insoluble solids from a scrubber liquor is to settle the solids by gravity. This can be achieved in a separate thickening tank, a remote pond or, for small installations, in the base of the scrubber. For large volumes of water, these sedimentation tanks occupy a great deal of ground-space and for this reason alternative thickening devices may be sought. Hydro-cyclones are relatively compact and perform well with larger solids, but have only a limited ability to deal with particles less than 30 to 40 microns in diameter. Centrifuges and filters may be feasible, but with dilute liquors their use may be precluded by high capital or operating costs. For the majority of applications therefore, even with its limitation of physical size, the thickener tank is chosen on the basis of its simplicity of operation, its flexibility to deal with varying loads and its relatively low cost.

Sedimentation thickeners may be batch or continuous in operation, both consisting of relatively shallow tanks from which the clear liquor is taken off the top and thickened liquor at the bottom (see Figure 13.1). To produce a clarified liquor the upward velocity of the liquor must at all times be less than the settling velocity of the particles. Thus, for a given throughput the clarifying capacity determines the tank area. The other key dimension of the thickener tank is the depth below the feed inlet. This governs the residence time for settling of particulates and hence controls the degree of thickening achieved. Residence times normally lie in the range of 1 to 4 hours. Flocculating agents help to increase the rate of sedimentation of some materials, particularly those likely to form colloidal suspensions.

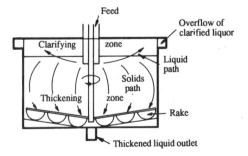

Figure 13.1   Flow in continuous thickener.

A thickener is normally sized on the basis of either past experience or laboratory sedimentation tests. Where the latter source is used, care needs to be exercised in sampling the liquors and applying the results.

Whatever the type of thickener, it is important to avoid unnecessary disturbance of the liquor. Gross overloading of the feed, air ingress or badly adjusted overflow weirs can all seriously upset the stability and performance of a thickener. However, the most common operational problem with thickeners is the reliability of the solids underflow system. The sizing and design of the discharge pipe, valve and pump must be carefully evaluated when it is required to discharge thick slurries. To avoid blockage problems, it is sometimes advisable to keep the sludge from over-compacting by continuously recycling the underflow liquor.

Some dusts may contain organic materials or substances which tend to form froths or scums. These constituents will most likely not sediment, but will float to the top of the thickener instead. To prevent such material being recycled with the overflow liquor, a baffle may need to be incorporated so that such contaminants can periodically be skimmed from the liquor surface. Continuous thickeners employ a rake or drag link conveyor to sweep sedimented solids towards the underflow outlet. The rake mechanism usually rotates at 0.1 to 1.0 rpm and is fitted with a torque limiting device to lift the rake if undue resistance is encountered. It is advisable to keep large lumps of solids such as slag or refractory from entering a continuous thickener. This can easily be achieved by fitting a small primary settler or coarse screen on the inlet to the main thickener.

Thickeners are usually designed to produce a clarified liquor with a solids concentration of not more than 1000 ppm w/w (0.1%). Better performance can be achieved but is not justified unless the residual dust is particularly hazardous to pumps, sprays, etc.

The form of the thickened liquor is very variable, depending on material properties more than the thickener design. Some materials agglomerate easily and pack down to form heavy thick slurries; others retain a high moisture level and are consequently wetter but easier to pump. The majority of inorganic substances thicken to a slurry consistency of 20% to 50% solids. Additional dewatering can be achieved with a pressure or vacuum filter, or a centrifuge, but the additional cost of this extra stage would need to be considered against potential benefits.

Even when the scrubbing liquor is recycled, fresh make-up water will be required to compensate for water lost with the sludge and due to evaporation.

It is advisable not to add make-up water directly into a thickener tank as this will disturb sedimentation and also prevent the full benefit of the fresh water being realised. It is better practice to allow the clarified water to overflow from the thickener into a recycle tank into which fresh water is admitted via a level controller. This same recycle tank can be used for chemical dosing when required.

14. FANS FOR GAS CLEANING PLANTS

The essential function of a fan is to provide the required flow against the resistance of the gas cleaning device and its associated ducting. Ideally, the fan should also be reliable and efficient in operation and be economic in capital cost. In practice, it may not always be possible to satisfy fully these secondary criteria.

A typical arrangement for the larger form of industrial fan is shown in Figure 14.1. Essentially, a fan has a rotating impeller carrying blades which exert a force on the air (or gas) thereby maintaining the flow and raising the total pressure.

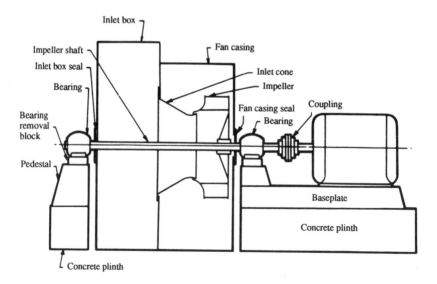

Figure 14.1   Typical arrangement for a large centrifugal fan.

14.1   TYPES OF FAN

There is a wide variety of forms of fan blades according to the application and duty required. The main classes are:

Centrifugal fans: forward-curved-blade;
radial-blade;
radial-tipped-blade;
backward-blade.

Axial flow fans: vane axial;
propeller.

When comparing fan types, the manufacturers will refer to criteria such as flow stability and fan efficiency. Flow stability is an indication of the ability of the fan to provide a fixed gas volume for a given pressure rise. An unstable fan will tend to 'hunt' between two displacement levels in a manner that is difficult for the operator to control. Fan efficiency provides a measure of how effectively the power provided at the fan shaft is converted into useful power in the gas stream.

Using these terms, Table 14.1 shows typical characteristics for each type of fan. It should be noted that, in this table, the efficiency levels refer to highly-engineered large industrial fans; smaller, cheaper fans may have efficiencies approximately 10% lower than the quoted values.

TABLE 14.1
Comparative characteristics of common types of fan

Fan type	Typical max efficiency %	Flow stability	Noise level	Applications
Forward curved blade	75	Doubtful	Moderate	Heating and ventilation: not suitable for dust
Radial blade	70	Good	Fairly high	Robust construction possible: suitable for high-dust loadings
Radial tipped	78	Good	Moderate	Widely used in USA; often described as self-cleaning
Backward blade	85	Good	Low	Widely used in dust applications
Axial	85	Doubtful	Higher than centrifugals	Not normally used for gas cleaning duties

Nowadays, the most widely used fan for gas cleaning applications is the backward-blade centrifugal fan. For severe applications, where the dust is sticky or erosive, the blades are either flat-plated, backward-inclined or radial. All backward-blade designs have the useful feature that the power curve is self-limiting, the power requirement reaching a peak at a flowrate somewhat higher than the normal selection point and thereafter decreasing with further increase of flow.

The static pressure/volume and absorbed power characteristics for various fan types are shown in Figure 14.2.

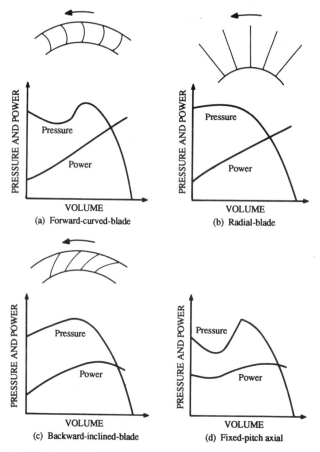

Figure 14.2 Static pressure/volume and absorbed power characteristics for various fan types.

14.2 FAN SELECTION

The type of information normally required for fan selection is as follows:

• volume required;

• fan suction and discharge pressures;

• gas condition — temperature, humidity, composition;

• characteristics of dust — concentration, stickiness, abrasiveness, toxicity, explosibility, pyrophoricity, hygroscopicity;

• type of drive — direct or belt-driven;

• method of control;

• space limitations;

• noise restrictions;

• efficiency required — to minimise motor size;

• corrosive application — special materials of construction, linings or coatings required.

The selection of fan size and speed is usually made from performance data published by the fan manufacturer. The user must always remember that a fan operating at a given speed can have an infinite number of operating points along the length of its characteristic curve varying from 'static no delivery' to 'free delivery'. However, when the fan is installed in a duct system, the operating point can only be that point where the system resistance curve intersects the fan characteristic curve, as indicated in Figure 14.3. Thus, in a given system, a fan operating at a given speed can only have a single rating which can be changed only by altering the fan speed or the system resistance itself.

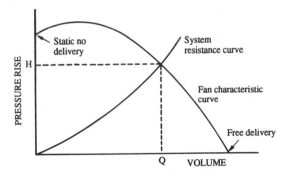

Figure 14.3 Single point of rating for a fan installed in a duct system.

119

TABLE 14.2
Fan laws for geometrically similar fans

General case	Q	$\propto$	ND^3	$\propto$	UD^2
	H	$\propto$	N^2D^2	$\propto$	U^2
	P	$\propto$	N^3D^5	$\propto$	U^3D^2
	efficiency = constant				
Diameter constant	Q	$\propto$	N		
	H	$\propto$	N^2		
	P	$\propto$	QH		
	P	$\propto$	N^3		
Rotational speed constant	Q	$\propto$	D^3		
	H	$\propto$	D^2		
	P	$\propto$	QH		
	P	$\propto$	D^5		
Tip speed constant	Q	$\propto$	D^2		
	H	=	constant		
	P	$\propto$	QH		
	P	$\propto$	D^2		

Key
Q = volume; H = pressure rise; P = power absorbed; N = rotational speed;
D = impeller diameter; U = impeller peripheral or tip speed
Any compatible system of units may be used.

14.3   FAN LAWS

The fan laws relate the performance of one fan to that of another, or to its own performance under different operating conditions. They are summarised in Table 14.2. The fan laws apply only to geometrically similar fans. In any particular design, all the important dimensions such as blade width, impeller inlet diameter and casing radii are proportional to the basic impeller diameter so that if this diameter is varied all the other related dimensions should be altered proportionally to ensure that the performance alteration is accurately predicted by the fan laws. In practice, impellers of differing diameters are fitted within the same casing and a reasonably close approximation to the 'fan law' performance prediction will be obtained if the variation in impeller diameter is small.

The most widely used fan law relates to speed variation where, as the fan is unchanged, the requirement for geometric similarity is fully satisfied. This is covered by the 'diameter constant' case on Table 14.2. In considering variation in fan performance, the effect of inlet density should be understood. A fan is a constant-volume machine but the pressure rise is proportional to the inlet density, as is also the power absorbed.

14.4   AERODYNAMIC SELECTION TO FULFIL PLANT VOLUME AND PRESSURE REQUIREMENTS

An essential requirement in fan selection is that the fan should achieve the pressure/volume characteristic offered by the fan supplier. A works test to demonstrate performance should be carried out in strict compliance with an agreed test code, BS 848[67] being appropriate in the UK. The class of volume and power tolerance to be satisfied, Class A or Class B, should be previously agreed. Where fan performance is critical to the satisfactory operation of the gas cleaning plant, Class A should be specified.

As there can be uncertainties in the specified design duty, the fan proposed should be chosen with a sufficient reserve in pressure and/or volume to accommodate some variation in design duty. A fan selected with a very high efficiency close to the peak of the pressure/volume curve may be unsatisfactory in practice as any increase in system resistance can result in unstable flow. Such considerations are even more important when two or more fans are required to work in parallel. In axial and forward-curved-blade centrifugal fans, the selection can be dominated by the need to achieve stability in operation.

A means of controlling fan output is required for start-up, and to provide output adjustment to follow variations in the process. The method of control chosen is also of great importance as regards stability of operation.

On centrifugal fans, three main control methods are usually employed.

DAMPER CONTROL

The damper is used as a simple throttle and can be remote from the fan. The fan operates according to its pressure/volume characteristic, the system resistance, and therefore the actual operating point, being varied by the amount of damper throttling.

Simple damper control is, however, wasteful of energy. When the fan is operating at reduced output, excess pressure is generated which simply dissipates across the damper. This may mean greater power consumption at lower volume than when the fan is at full rated output, particularly with backward-inclined-blade fans.

INLET VANE CONTROL

Inlet vane control is particularly useful when the system resistance does not follow a square law behaviour; even a constant pressure system can be controlled in a stable manner over wide volume variations.

Movable vanes in the immediate inlet to the fan impeller are used to impart a swirl to the gas entering the impeller, the intensity of swirl increasing as the vanes are closed. The effect is markedly different from simple damper control as the 'throttling' effect is absent; each inlet vane position gives different pressure/volume characteristics which become progressively steeper as the vane closure increases.

Using inlet vane control assures stability at duties below the design duty. Even in fault conditions, with substantial pressure increases above design, any loss of stable operation can be overcome by partial closing of the vane control.

Other control devices which produce a swirl effect, such as inlet louvre dampers in the fan inlet box, can give similar results as regards stability, but the power savings at part load are not so great.

SPEED CONTROL

This can be a most effective method of control particularly when the system resistance follows a square law behaviour. For each speed setting of the fan, a different pressure/volume and power characteristic applies, so that the operating condition maintains a constant relationship with the surge point. If the variable speed drive had a constant efficiency, then the overall efficiency would be constant; however, many variable speed drives, electric motors, fluid couplings and steam turbines themselves have efficiencies which vary with speed, and this should be considered when making an assessment. An important advantage of variable speed control is that both noise level and rates of wear reduce significantly at lower speed. A disadvantage may be the higher cost and mechanical complexity of this device.

Figure 14.4 shows the effects of the various control methods on the basic fan pressure/volume and absorbed power characteristics. The surge lines shown indicate the normally accepted limits of stable operation. Operating points to the left of the surge line are best avoided except for low pressure fans operating as single units.

Before leaving stability and control, it should be borne in mind that bad flow conditions, outwardly evidenced by ductwork pulsations, panel and stif-

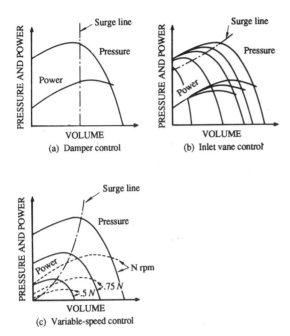

Figure 14.4   Fan control methods.

fener failures, etc, are not necessarily caused by fan instability. Poor ductwork design, with flow breakaway, turbulence and even reversal could be the cause. Analytical design of complicated ductwork systems is difficult and a practical approach using flow models may be worthwhile.

14.5   RELIABILITY IN OPERATION

In continuous process plant reliability is of paramount importance, particularly where loss of the gas cleaning system will force a stoppage of the whole plant. Good basic engineering in the fan is essential. Gas cleaning duties can be arduous and the design and construction of the fan should recognise this fact. Whatever fan design is used, it is fundamental that the shaft and bearings are well engineered. In-service out-of-balance is to be expected and margins in bearing capacity and in shaft stiffness are necessary to avoid damage.

Foundation design can affect the smooth running of a fan. In an extreme case, a flexible foundation could initiate fan shaft whirl to the extent that a shaft failure could result. With a floor-mounted fan, the fan supplier would not normally be responsible for the design or supply of the foundation, but values of his required vertical and transverse stiffness relative to the bearing mountings should be obtained so that the civil designer has a correct basis for his work. The foundations should, of course, be adequate in design and construction to ensure that the various parts of the fan and motor assembly are maintained in good alignment. A check should also be made to ensure that the natural frequency of the whole, or any part, of the foundation system is well separated from the rotation frequency and its harmonics. Where a fan is to be mounted on steelwork, either directly or on anti-vibration mountings, very careful design is essential to ensure alignments are maintained and kept free from harmful resonances.

Being stationary, the fan casing is not so critical, but the construction will be dictated by the amount of erosion or corrosion to be expected. If only a light burden or non-corrosive dust is to be handled a normally robust casing will suffice, but if the dust burden is heavy the fitting of replaceable wear liners is necessary. If corrosion is the main problem then special coatings, from chemical resistance paint through to hard vulcanised rubber, can be used.

The fan impeller design will largely be dictated by the service conditions. Where severe erosion is expected an impeller design capable of carrying replaceable wear liners should be used. Radial blades, radial-tipped and both backward-flat and backward-curved blade designs can be so protected. The radial-blade and backward-flat-blade designs are geometrically more simple and are consequently easier to protect and maintain. Where erosion is less severe, robust, unlined, radial-tipped, backward-flat and backward-curved blade fans can be satisfactory.

There will be a limit to impeller life, but first cost considerations may make such an approach economic. The rate of fan erosion is approximately linearly related to the magnitude of the dust burden and to the peripheral speed following a law generally considered to lie between a square and a cube function. (Rotational speed in itself does not control wear to any significant extent.) It follows that high burdens at low peripheral speeds can be much less arduous, as regards erosion, than low burdens at high peripheral speeds. Corrosion can be resisted by coatings, but if erosion is also present the coating life may be limited. Where impellers must resist corrosive attack the most satisfactory solution is to use special materials such as stainless steels, aluminium, bronze, titanium, inconel, etc. Some such materials have mechanical strength limitations, or present fabrication difficulties, and the fan design may be dictated by these conditions.

14.6   NOISE

Fans are basically noisy machines and it is increasingly necessary to control their noise emission to ensure acceptable working conditions within the plant and to avoid causing a noise nuisance in the surrounding area.

The noise produced by a fan is a combination of broadband noise and interaction noise. Broadband noise covers a wide frequency range and is associated with random vortex shedding at surfaces and in passages. This source of noise cannot be eliminated, but it is somewhat reduced where the aerodynamic efficiency is high. Interaction noise occurs at blade-passing frequency and its harmonics. On low-speed fans, the interaction noise is frequently swamped by the broadband noise, but as speed increases the interaction noise increases at a greater rate than broadband and can become dominant. Interaction noise is near to a pure tone and can be very noticeable and objectionable. Alterations to the detailed design, such as increasing the clearance between blade tip and casing on a centrifugal fan, can reduce this interaction noise, but the amount of reduction possible is limited as such changes affect the fan performance.

The general level of noise generated by a fan varies with both the diameter and speed, with increase in speed having a slightly greater effect than increase in diameter. Figure 14.5 overleaf shows typical noise spectra for three fans where the volume is constant and the pressure varied.

The pressure rise obtained from a fan is, by the fan laws, a function of tip speed, ie the product of diameter and speed. Figure 14.5 clearly shows how increased noise level is linked to increased pressure rise. The increased pressure rise in the example has been obtained basically by increasing speed and the pronounced increase in interaction noise at blade passing frequency is clearly shown.

The fan supplier will normally specify the fan noise as the sound pressure level that would be measured at various frequencies and at a stated distance, usually 1 metre, from the fan. A weighted 'A' scale average would also be commonly stated. 'A' scale weighting gives an average sound pressure level which approximates to what the human ear hears. At frequency levels below 500 Hz, the spectrum values are reduced as the ear is less sensitive to low-frequency noise.

If the fan is fully ducted the noise at, say, 1 metre will be the 'in-duct' fan noise less the reduction given by the casing material. The casing reduction increases with thickness but according to a law of diminishing returns. If very high attenuations are demanded, acoustic lagging will be required.

125

Fan	Pressure rise cm wg.	Diameter m	Speed rpm	Blade passing frequency
A	25	1.35	980	196 Hz
B	60	1.25	1480	296 Hz
C	152	1.00	2970	594 Hz

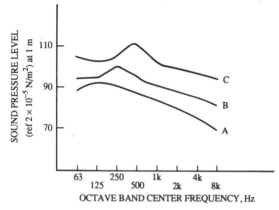

Figure 14.5   Typical noise spectra as a function of fan pressure rise.

Noise levels are normally quoted as measured in 'free-field conditions'. Such values can only give a general indication to the likely levels on a confined site where other noise sources and reflecting surfaces are present. There should also be an awareness that vibration and resonance in ductwork can sometimes magnify the noise levels to a marked degree.

Attention must also be paid to fan shaft sealing and the tightness of the casing joints, damper spindles, etc. Noise is air-borne and any leakage path will allow noise to radiate from the point of leakage.

In a fully ducted installation, it should be remembered that the connecting ductwork and any expansion joints will require the same consideration as the fan itself if noise break-out elsewhere is to be avoided.

To completely specify the noise generated by a fan the in-duct noise levels, or open inlet and discharge levels, should also be specified. This will provide the necessary basis for silencer design, or for calculation of end noise at extraction hoods or at the discharge from stacks.

In considering the total noise in a fan installation it should not be forgotten that there will be a noise contribution from ancillary items such as driving motors, gearboxes, etc.

When deciding on what measures are necessary to control noise from fan installations, specialist advice should be sought. In some installations, legislative noise levels can be achieved and still prove unacceptable to employees, neighbours, or both. If, for example, a fan installation is adjacent to office accommodation, a lower level of noise may be required to prevent a nuisance or interference with speech. Finally, the legal rights of neighbours (see Chapter 3) should not be overlooked.

Normally silencers are of the absorption type and especially with wet or very dirty gases the infill may have to be protected and be removable for cleaning. Reactive-type silencers may be incorporated in systems where dust build-up is particularly pronounced but on wet systems it is important that the silencer is not allowed to accumulate water or dust as this will alter the silencing characteristics.

Acoustic control is expensive and if, for example, in-line silencers are fitted, the extra power required to overcome their pressure drop will be a continuing extra running cost throughout the life of the plant. If acoustic controls are added other than at the time of original installation the cost is likely to be greater and any additional pressure drop incurred could result in a reduction in the volumetric capacity of the extraction system.

14.7   FAN DRIVES

The majority of fans are driven by electric motors, mostly constant speed, 3-phase, squirrel-cage, induction machines. Such motors are simple, robust and reliable, but they can suffer from difficulties in starting the fan, associated with both the motor and the power supply.

Fans are inherently high-inertia machines and to ensure good starting the motor supplier must be advised of the rotating inertia of the fan, the fan bearing break-away torque and the absorbed power when running at the design speed. If dampers or inlet vanes are fitted the power at speed with these closed should also be stated, as this reduction in impeller absorbed power will make more torque available for acceleration. For hot gas fans, some form of fan control is usually essential to avoid an excessive margin to cover the increased power on cold start-up. All this information will enable the motor supplier to assess the starting problem and decide on the motor and starter required.

The simplest and most reliable starting arrangement is a direct-on-line system. Where the fan inertia is high relative to the power absorbed or where the electrical supply is inadequate other starting systems may be necessary. Star/delta, auto-transformer and slip ring motors with external resistances can all be used.

It can sometimes be an advantage to use a synchronous motor, particularly if the fan driving power is substantial. By adjustment of this motor's excitation system the supply system power factor can be corrected, so reducing overall power costs.

Variable-speed motors can be very attractive from the point of view of power saving if the fan cycle is variable. They are expensive, however, and a careful assessment of expected savings against increased cost should be carried out.

Variable-speed is also obtainable by using a constant speed motor driving through a fluid coupling. Speed variation is obtained by controlling the degree of filling of the coupling, and although there are slip losses this is an effective and reasonably efficient drive. The starting characteristics are excellent as the motor can be run up to full speed, virtually unloaded, before accelerating the load in a controlled manner.

If steam is available, the turbine drive is worth some consideration. The turbine can be variable speed if required and despite the relatively low efficiency of small turbines the operating costs can be attractive.

14.8   IN-SERVICE REQUIREMENTS

Positive action is required to ensure that a fan will continue to operate in a satisfactory manner after installation. Regular monitoring of the state of the bearings and the rotational balance of the impeller will give a good indication of the fan's mechanical condition and help to avoid unexpected breakdown.

The degree of instrumentation required will depend on the application and the fan size. A low flow alarm is often of value, and for larger fans it is advisable to monitor bearing temperatures. Continuous measurement of the vibration level with a presentation of actual amplitudes of vibration enables any deterioration to be identified and checked at an early stage.

Local indication may be sufficient on small installations but if there is a central control point the information should also be displayed there. In addition to the actual measurements, limit switches for 'warning' and 'stop' should be included.

On critical installations, instrumentation should be included to indicate motor amps, fan control position, fan pressure, etc at the control station, and where pressure lubrication bearings are employed, bearing oil pressures, tank oil levels, etc should also be displayed.

Fan suppliers normally provide a manual covering the operation and maintenance of the equipment. The user should ensure that guidance on all aspects of operation and maintenance is given for his particular application. For example, a fairly high wear rate may be anticipated and the manufacturer's instruction should clearly state the checks and measurements necessary to ensure that the acceptable and safe limit of use is not exceeded. The 'Health and Safety at Work' legislation requires that both the supplier and the user ensure that the operators have adequate guidance available to them.

15. INSTRUMENTATION AND CONTROL

Ideally, good instrumentation should assist plant operators in getting the best out of their dust control plant by monitoring key parameters and giving early warning of non-optimum performance. On the other hand, poorly selected or maintained instrumentation will cause considerable aggravation and may reduce the effectiveness of pollution control, even though there may be no fault in the dust collector itself. It will not be an easy matter to achieve the former result and avoid the latter.

By their very nature, dusty atmospheres are liable to cause problems to many types of sensor that are successfully used in clean-air conditions. Dust ingress can plug, absorb, corrode or insulate sensitive elements so that their apparent output can be quite unrelated to the true conditions of their environment. Thus, from the outset, the user must employ great care in selecting and relying on instruments which will work in association with gas-cleaning plant. Even well-suited instruments will almost certainly require regular expert maintenance and calibration. If it is not practical to allocate this degree of care, the investment in instrumentation will be wasteful.

Having given this salutary warning, it is evident that a few carefully selected and maintained instruments are generally a better recommendation than many superfluous indicators that do little more than add unnecessary expense. It also follows that key indicators or recorders should be located where they can easily be seen by the operator responsible for supervision of the gas cleaning plant, and that individual sensors and control elements should have appropriate access points for inspection and maintenance.

As a minimum requirement for most gas cleaning systems, the pressure drop across the main arrestment unit should be measured. Gas inlet temperature and flow are often necessary data, and it is frequently useful to install an ammeter showing the current drawn by the main fan. These parameters can be indicated, recorded and alarmed at critical settings, or used for closed-loop control as, for example, on gas flow regulation and temperature protection of bag filters. Where possible, the instruments should also be sited so as to avoid obvious areas of dust or moisture accumulation, although any projection into a dusty gas stream will suffer dust impaction with time. It is also necessary to consider striation effects when siting instruments downstream of duct bends or changes in cross section. Where practical, it is better to locate the sensor in flow areas which are aerodynamically stable and not subjected to rapidly changing conditions of temperature or concentration gradient.

When calibrating instruments, a little extra time spent in cross-checking values from two independent sources can often be well worthwhile. For example, a gas flow meter could be checked by a pitot survey and then compared with the manufacturer's curve of the displacement produced by the installed fan. Additionally, it should then be possible to relate the fan power requirement to its drive motor current. The advantage of this technique is that the operator is much better able to judge whether subsequent variations in the reading of an indicator show a true change in process conditions or merely a drift in the instrument.

Within the context of this guide, it is not possible to describe in detail the wide range of general-purpose instruments that can be applied to dust control. However, one area of increasing importance to operators of dust and fume control equipment concerns the automatic monitoring of particulate emissions from vents which should, ideally, give plant operators and control agencies a continuous indication and record of the effectiveness of gas cleaning. A number of devices are marketed which measure the obscuration of a light beam projected across a duct and there are others which attempt to relate dust concentration to back-scattered light. Other methods include extracting a single point sample from a duct, collecting the solids on a tape filter and estimating the solids loading by pressure drop or B-ray absorption. With an adequately maintained instrument carefully selected and sited according to the particle size distribution and nature of the solids, a useful qualitative indication of a marked change in collection efficiency can often be obtained. Because of their sensitivity to such factors as particle size distribution, shape, density and other process variables, correlation of the instruments to give reliable concentrations or mass emission read-outs is extremely difficult. Inexperienced users should view with scepticism claims that specialised equipment can provide this service. Practical experience of most installations shows that reliable quantitative measurements can only be achieved through the sampling techniques outlined in Chapter 18.

16. DISCHARGE OF EXHAUST GASES

Having worked to contain, convey and collect the offending dust or fume, it is all too easy to discharge the cleaned gas carelessly and later regret the lack of attention to what would appear to be a simple, final step. There are a number of factors that must be considered.

First, will the exhaust vent satisfy statutory requirements? Under Section 6 of the Environmental Protection Act 1990, an operator of a 'prescribed process' is required to obtain an authorisation for emissions, including those into the atmosphere, from the appropriate regulatory authority. Processes are prescribed by the Secretary of State in regulations[68] and include many activities carried out in the chemical and process industries as well as other industry sectors.

The regulatory authority in the case of processes which are recognised as being potentially more polluting is Her Majesty's Inspectorate of Pollution (HMIP), which operates a system of Integrated Pollution Control (IPC). IPC considers discharges from industrial processes to all environmental media in the context of the effect on the environment as a whole. Dust, fume and noxious gases arising from these processes are subject to the conditions of IPC authorisations. Less potentially polluting processes are regulated for air pollution only by local authorities and their environmental health departments.

Under Section 7(4) of the Environmental Protection Act, an authorisation will contain conditions requiring the operator to use 'the best available techniques not entailing excessive cost:

(a) for preventing the release of substances prescribed for any environmental medium into that medium or, where that is not practicable by such means, for reducing the release of such substances to a minimum and for rendering harmless any such substances which are so released; and

(b) for rendering harmless any other substances which might cause harm if released into any environmental medium.'

Exhausts to atmosphere from other processes, where nuisance and disamenity are the principal concerns, are subject to planning regulations by local authorities and control by their environmental health departments. Assuming that the arrestment equipment has been adequately designed to achieve acceptable particulate concentrations in the exhaust gases, the control authorities will want to confirm that discharge stacks are of a sufficient height and suitably located to ensure that residual pollution will be adequately dispersed to avoid excessive ground-level concentrations.

The subject of dispersion has been extensively researched with complex formulae and computer programs developed to model pollution patterns resulting from stack emissions[69, 70]. However the results from such studies lack precision and need expert interpretation.

As a rough guide, the following formula can be used:

$$H^2 = \frac{9M}{20C}$$

where H =   stack height (metres)
        M =   mass rate of emission of pollutant (kg per day)
        C =   maximum permitted 3-minute mean ground level concentration (mg/m^3). C as a first approximation can often be taken as one fortieth of the Occupational Exposure Limit for the pollutant (see Chapter 3).

As a general rule, minimum chimney heights should be 5 m above roof ridge level and adjacent buildings. Adjustments must be made for such factors as thermal plume rise, efflux momentum, topography, including nearby buildings, and the type of area in which the factory is situated. Expert advice should be sought from the control authorities.

The efflux velocity should be appropriate to the type of material being discharged. For dry discharge, it is important that the velocity should be sufficient to avoid air ingress to the stack and downwash, where the plume travels down the outside of the stack in the wake of the prevailing wind. Both ingress and downwash can lead to poor dispersion and damage to the stack top. In general, a velocity greater than 15 m/s is sufficient to avoid these problems. Coned top sections can be used to achieve this condition, but fan power should be available to cope with the restriction and positive pressure sections in the stack must be acceptable.

With wet exhausts, discharge of droplets must be avoided since a slight solids content can cause severe spotting. Even after efficient droplet disentrainment, excessive gas velocity in the stack will cause a liquid film on the chimney wall to creep up the stack, breaking up to droplets within the stack or at the discharge lip. Normally, gas velocities less than 9 m/s in the barrel of the chimney will guard against this difficulty. Effective stack drainage must also be provided.

Proper provision must be made for sampling stack gases. This requirement is essential for testing equipment against supplier guarantees and emission control limits, and is highly desirable for routine monitoring of the system performance. A straight length of duct of at least ten equivalent diameters should be designed into the exhaust system so that accurate velocities can be determined and representative dust samples taken. The sampling points should, of course, be readily accessible. An adequately-sized platform should be provided for testing which should, of course, be safe in all weathers. Care must be taken to avoid excessive swirl in the duct at sampling points such as after inertial droplet separators on wet washer systems.

Finally, to assist dispersion and avoid pressure areas in the stack, particularly at sampling points, the discharge vent should not be obstructed with devices such as cowls. Protection against rain ingress can be designed to avoid such features[15].

17. ECONOMIC EVALUATION OF PLANT

Responsible engineers will want to ensure that in selecting dust and fume control plant they make a choice which is both technically and economically sound. The selection procedure should initially concentrate most heavily on the technical merits of the competing processes. As the selection procedure develops, however, it is not only proper but essential that economic factors should play an increasing role. Economics cannot make a technically bad solution acceptable, but it should influence the choice between alternatives that are otherwise equally satisfactory in terms of the primary selection factors. Thus, it is envisaged that the reader may use the selection procedure to effect his first choice and then find that the cost of this technically best solution is unacceptable. He may then find it necessary to repeat the selection procedure choosing an acceptable, but possibly less desirable, route to determine if the overall cost of that possibility is more favourable. At the end of the day, a degree of compromise may be necessary, but it is of paramount importance that the pollution control plant meets its main objectives. If it fails to do that not only has money been wasted, but the operation of the main process may have been put at risk of forced closure.

Whilst it is true to say that cyclones and fibrous filters are often associated with small, low cost projects, that wet dedusters and bag filters are often selected for medium sized contracts and that electrical precipitators are often applied to very large schemes, such a simple guideline can be unfavourably applied and in some cases can be totally misleading. Certainly, between wet dedusters, bag filters and electrical precipitators the reader would be ill-advised to assume any cost preference until he has at least conducted some budgetary cost comparisons for the total supply of collector and all major ancillaries for each case.

Even when estimating the provisional cost of plant it will usually be necessary to consider the total equipment supply, not just the dust collection device. In more and more cases, it is found that the cost of ancillaries may be several times the cost of the dust collector, so comparing isolated cost items may be highly misleading. Again, it cannot be assumed that different dust collectors will require the same ancillaries. The ducting for a high-temperature collector may be totally different from the ducts required if the gases are pre-cooled. A dust collector operating with a high pressure drop will want a very different fan from that required with a low-resistance collector. Thus, whilst economic short cuts are sometimes inevitable, they need to be used with extreme caution.

In assessing the capital cost of a dust or fume control project, Table 17.1 can be used as a check list of components that may need to be included within a scheme. Not all the items listed may be necessary for a particular application, but if they are wanted then their cost should be budgeted from the outset. At later stages in the assessment some of the non-essential items may be left out, but over-reduction may prejudice the satisfactory operation of the scheme as a whole.

By themselves, budget capital costs will not provide sufficient information to make a rational judgement between alternative schemes. Instead, their total should be regarded as a partial contribution to the cost of operating various schemes over the full projected working life of the plant. This choice of operating life may be important; the longer the period, the more it will favour plants with

TABLE 17.1
Checklist of hardware items

Hoods or enclosures
Ducting
Control dampers
Dust collector, complete with hoppers
Fan
Stack
Instrumentation
Structural steelwork
Access provision for operation, maintenance, testing
Foundations and civil engineering
Electrics and control wiring
Dust/slurry removal from hoppers
Dust/slurry treatment plant
Water treatment/recycle plant
Dust/slurry conveyors
Dust/slurry storage/disposal plant
Services supply plant (air, steam, water, etc)
Recommended spares for the above

Note: All costs to include design, manufacture, transport, erection, commissioning and testing.

TABLE 17.2
Checklist for cost comparisons

Annualised capital cost	Depreciation Financing
Annualised operating cost	Manual and skilled labour Maintenance: labour, materials, equipment Services: electric power, fuels, water, steam, compressed air, chemical additives Dust/slurry disposal costs
Annualised overhead costs	Plant management and administration Staff facilities Laboratory support Insurance Training Rates and taxes

Note: The above annual costs can be used to construct both a cash flow prediction and a total
cost over a given number of years of plant life.
For rigorous calculation discounted cash flow or similar techniques should be used.

low operating costs. Thus, a low capital cost plant may prove expensive over a total of, say, 15 years if its power cost is an order of magnitude higher than that of an alternative plant of much higher capital cost. To effect this comparison rigorously it is necessary to use one of the widely accepted techniques, such as discounted cash flow, which take into account factors such as the time-value of money and inflation allowances[72].

For initial assessments, however, it is normally acceptable to use present day fixed costs to determine which schemes warrant a detailed examination. For this purpose, Table 17.2 indicates some of the factors that should be included in assessing a total cost commitment. The list may not be all-inclusive and the user should include any additional costs associated with his particular application.

When a collector type has been selected, a more detailed evaluation of equipment will need to be undertaken. The purchaser can obtain[71] a comprehensive checklist of the various components for each main collector type which will help in drawing up a full specification to send to suppliers. This should enable a full assessment to be carried out before the final decision to purchase is made.

18. SAMPLING AND TESTING DUST-LADEN GASES

This chapter deals with dust sampling both from ducted gas streams and from the atmosphere. Sampling from the atmosphere is carried out mainly to monitor health hazards and the techniques involved are usually quite different from those used for sampling gas streams. In either case, in the course of sampling, it will be necessary to measure accurately quantities such as gas flowrates, temperatures, humidities, pressures, etc. Sometimes a chemical analysis of the gas sample will also be required. Physical measurements, such as particle size analyses, will often be required on the dust sample.

Sampling and testing will frequently be carried out on a pilot plant in order to obtain valuable design data for different or new applications. Sampling will also be carried out as part of a performance test following the start-up of a new plant.

18.1 DEFINITION OF GAS CONDITION

When discussing gas volumetric flowrates, it is essential always to state the temperature and pressure of the gas. For most practical purposes, the volume of a fixed mass of gas or air is inversely proportional to pressure and directly proportional to temperature, measured in absolute units. When gas flowrates are given at the actual pressure and temperature of the gas, this is termed the 'actual' gas flowrate and the abbreviation 'a' is used (ie am^3/s means actual cubic metres per second).

It is often convenient, however, to specify the gas flowrate at a standard set of conditions, which would represent the flowrate of the same mass of gas if its temperature and pressure were changed to the standard levels. The standard conditions which are generally used in the dust and fume control industry are referred to as 'Normal' conditions which are 0°C and 760 mm Hg pressure, and the abbreviation 'N' or 'n' is used (ie Nm^3/s means normal cubic metres per second). Historically, the term 'standard' referred to conditions either at 15.6°C or 20°C and 760 mm Hg pressure, but these conditions are becoming less widely used.

The advantage of working at normal conditions is that the gas volume can be considered as remaining constant throughout the process even when the temperature and pressure change; also, when two or more gas streams at different temperatures, say, join together, the total volume (at normal conditions) is the sum of the individual volumes. It should, however, be noted that the above standard is not used in all industries and where there is likely to be doubt, the gas temperature and pressure should be stated instead (eg 1500 m³/s at 0°C and 760 mm Hg pressure).

In stipulating gas flowrates, it is also important to make clear whether or not the moisture content is included. It is often convenient to consider the dry gas separately in situations where the moisture content varies due to changes in operating conditions.

18.2   MEASUREMENT OF GAS FLOWRATE

Measurement of gas flowrate is achieved usually by measuring the differential pressure head across an orifice plate, a venturi meter or a pitot-static tube[74, 79, 81]. Rotameters are also used, particularly with dust sampling equipment.

The orifice plate is a permanent installation which requires straight runs of ducting of at least nine diameters (six upstream and three downstream). The significant pressure drop across the orifice plate often makes it unattractive for gas cleaning and ventilation applications, although it is widely used with gas sampling equipment. The venturi meter, on the other hand, operates with a lower pressure drop than an orifice plate, but it is more expensive.

The pitot-static tube is the flow-measuring device most widely used in dust and fume control. It is not a permanent installation but is inserted into the duct when measurements are required. The device consists of two concentric tubes which, at a short distance from one end, are bent at right angles into the gas stream. The inner tube is open at the end and measures the total or impact pressure of the gas stream. The end of the outer concentric tube is sealed and a series of small orifices on the outer surface a short distance from the head of the instrument give an accurate indication of the static pressure. The other ends of the concentric tubes are connected to a manometer to measure the difference between the impact and static pressures, ie the velocity pressure, which is directly proportional to the kinetic energy of the fluid, thereby allowing the gas velocity to be calculated. Because the gas velocity is proportional to the square root of the pressure difference, pitot tubes become insensitive at low gas flowrates.

Pitot tube readings must be made at several positions across the duct and from these point velocities the average velocity, and hence the total flowrate, are easily calculated. Guidance is given in BS 1042[81] on the number and position of the readings required. The pitot survey should be carried out in a straight length of ducting, if possible, to minimise the effect of turbulent eddies on the accuracy of the pitot readings.

Other velocity measuring devices are available, but they have limited applications in pollution control work. For example, vane anemometers are suitable only for non-toxic, non-corrosive gases at ambient temperatures and low velocities. Similarly hot-wire anemometers are suitable for low gas velocities and only in clean conditions.

18.3   MEASUREMENT OF GAS TEMPERATURE

Accurate definition of temperature conditions in conjunction with gas flowrates is essential for correct sizing and selection of gas cleaning plant.

When measuring gas temperatures care must be taken to avoid radiation heating effects to or from the duct walls or from exposed heat sources.

Thermocouples are the most widely used temperature measuring device and the most common thermocouple materials are:

- Copper/constantan   from −200°C to 400°C
- Iron/constantan   from 0°C to 800°C
- Chromel/alumel   from 0°C to 1100°C
- Platinum/platinum-rhodium   from 0°C to 1600°C

Other materials are available for temperatures up to 2500°C.

When used in corrosive atmospheres, thermocouples must be sheathed for protection. Below 1000°C metal sheaths are satisfactory and re-crystallised alumina sheaths can be used up to about 1850°C. Allowance must be made for the temperature drop across the sheath and possible dust build-up on the surface.

The following reference points are generally used for calibration of temperature sensing elements:

- Melting point of ice   0°C
- Boiling point of water   100°C
- Boiling point of sulphur   444.6°C
- Melting point of silver   960.5°C
- Melting point of palladium   1552°C

Mercury-glass thermometers, covering a range from below 0°C to 350°C, are widely used and alcohol-in-glass thermometers extend the range down to −80°C.

Care must be taken that the thermometer is positioned in a representative position in the gas stream and with large ducts this becomes very difficult.

SUCTION PYROMETER

A useful method of avoiding radiation effects is to shield the temperature sensing element from any radiation surrounding it with a 'radiation shield' and aspirating the gas past the element at high velocities to increase the convective heat transfer.

The accuracy of these devices depends on the actual gas temperature of the surroundings and the velocity of the gas. Further details of this type of equipment and the necessary correction factors can be found in Reference 25.

ESTIMATION OF DEWPOINT

The temperature to which a particular sample of air or gas has to be cooled before depositing moisture droplets is termed the dewpoint and is a direct measure of the moisture content of the gas. However, the presence of even very small quantities of sulphur trioxide can greatly elevate the effective dewpoint above that indicated by the moisture content alone, and can lead to serious corrosion problems if not detected.

Dewpoint data can be obtained by:

• calculation from the quantity of moisture in the air or gas entering the process plus that produced by the process (for example by combustion) and making suitable correction for sulphur trioxide content;

• measurement using wet- and dry-bulb thermometers;

• measurement using automatic dewpoint measuring techniques.

18.4 MEASUREMENT OF PRESSURE AND PRESSURE DIFFERENCES

Pressure measurements are frequently required to determine resistances across hoods, ducting, collectors, etc or pressure differences across fans or flow-measuring instruments. When measuring static pressures in a duct, the reading should be carried out at a point where the gas velocity is parallel with the duct wall. The static pressure opening is usually a simple hole (1.5–3.0 mm diameter), drilled flush with the inner surface of the wall. It is important that there are no burrs or projections on the inner surface. Alternatively, the static tube of the pitot may be used.

In dust and fume control systems, pressure (and pressure drops) are generally very low and manometers are widely used. The simplest type of pressure gauge is the vertical U-tube manometer which is usually calibrated in cm water gauge. Various manometric liquids are used depending upon the pressure range required; these include water, alcohol, oil, mercury, kerosene, etc. For easier reading of the instrument, one leg of the U-tube is often replaced by a reservoir or well (well-type manometer) so that only one limb of the manometer requires to be read.

Increased accuracy and/or lower pressure readings can be achieved by tilting one leg of the U-tube to form an inclined manometer or draught gauge. Again, very often only the inclined leg is used with the other leg replaced by a reservoir. The accuracy of the gauge is clearly dependent on the slope of the inclined tube. There must, therefore, be provision for levelling the instrument and for zeroing the scale. The inclined manometer is used widely with the pitot tube.

A modification of the inclined manometer is the inclined-vertical gauge in which the indicator leg is bent to give both a vertical and an inclined section. The advantage over the simple inclined manometer is the smaller physical size for a given range whilst retaining the refined measurement afforded by the inclined section.

There are many other methods of pressure measurement available, based on the use of bellows, diaphragms, liquid-sealed bells, transducers, etc. Aneroid-type gauges are widely used, the best known example of which is the Magnehelic gauge.

18.5 COLLECTION AND ANALYSIS OF A GAS SAMPLE

If the bulk of the gas stream is air then it will not generally be necessary to analyse the gas for its constituents. If an analysis is required then care must be taken to draw a representative sample from the bulk gas flow and to avoid any stagnant zones. Heated sample probes may be required as care must be taken to avoid inleakage of air or condensation within the sample lines caused by rapid cooling.

Analysis of gas samples is usually a sophisticated procedure which can be obtained through a commercial laboratory if the techniques are not available to the user.

An 'Orsat' apparatus can be used quite effectively on site for measuring the quantities of carbon dioxide, oxygen, carbon monoxide and nitrogen in a gas sample. This method simply uses different liquids in turn to wash a specific volume of the test gas. Each liquid absorbs one of the reactive constituents in the gas and the quantity of that constituent is estimated by the reduction in the original volume of gas. The quantity of nitrogen is assumed to be the volume of sample gas left after all the washing procedures have been completed.

A useful method of measuring the concentration of certain gaseous constituents is by means of gas detector tubes. These are thin glass tubes containing crystals which change colour when exposed to certain gaseous chemicals. A prescribed quantity of the gas to be tested is drawn through the tube (by means of a hand-operated bellows pump) and the extent of the colour change indicates the concentration of that particular constituent. Detector tubes are available for a range of chemicals including sulphur dioxide, oxygen, carbon monoxide, carbon dioxide and nitrogen oxides.

18.6   DUST SAMPLING FROM A DUCTED GAS STREAM

The best method of obtaining a representative dust sample is to draw off a small fraction of the gas stream through an aspirated nozzle and probe tube and collect the dust on a filter. If the mass of dust captured and the quantity of gas passing through the filter are measured, then the dust concentration in the gas stream can be obtained. A typical sampling train is shown in Figure 18.1 overleaf. The sampling nozzle inserted in the gas stream is pointed upstream and, if required, the sampling probe and filter are heated to prevent condensation. There are many different types of filters and separators available for collecting the dust sample; double-layer glass-fibre or glass-wool filters are widely used except perhaps where a particle size analysis is required when a cascade impactor might be used. Catchpots are often incorporated in the sampling train to cool the gas and to enable its moisture content to be obtained by weighing before and after a test. The gas volume passing through the apparatus can be measured by a gas meter, an orifice plate or a rotameter.

An alternative arrangement to that shown in Figure 18.1 is provided by the BCURA apparatus[74] which uses a small-diameter, high-efficiency cyclone for the dual purpose of measuring the flowrate of the gas sample and collecting the dust in a small hopper at the base. Any dust passing through the cyclone is captured by a glass-wool filter. The cyclone and filter assembly is generally used inside the duct with the sampling nozzle screwed directly on to the inlet of the cyclone; alternatively, the assembly can be used outside the duct, but this normally requires a separate flowmeter for measuring the volume of gas sampled.

Whatever sampling technique is used, it is important that the gas sampling rate is adjusted so that the velocity of the gas drawn into the nozzle tip is the same as the bulk gas velocity in the vicinity of the nozzle. Otherwise, the inertia of the dust particles travelling in the bulk gas stream will lead to apparently low dust loadings if the sampling velocity is too high, and to apparently high dust loadings if the sampling velocity is too low. This principle of 'isokinetic' sampling is illustrated in Figure 18.2.

Because the gas velocity and dust concentration are likely to vary across the duct cross section, it is necessary to take samples from a number of different positions. The sampling positions are determined by dividing the cross-sectional area into smaller equal areas and samples are taken isokinetically at the centres of gravity of the sub-areas. The accepted method of sampling, including the recommended number and position of the sampling points, is given

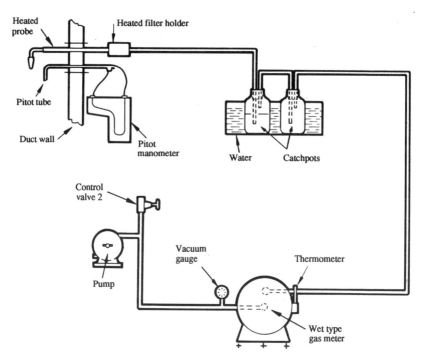

Figure 18.1   Diagrammatic arangement of typical dust sampling apparatus.

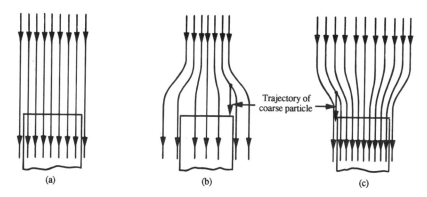

Figure 18.2   Principle of isokinetic sampling. (a) Isokinetic sampling — correct concentration and particle size distribution; (b) Sampling velocity too low — excess of coarse particles; (c) Samplig velocity too high — deficiency of coarse particles.

in BS 893[80]. The method is lengthy and complicated and for more rapid investigations, a simplified procedure is given in BS 3405[85]. Further information on the different techniques used is given in References 73–77.

Full details of how and from where dust samples are obtained should always be given to ensure that it is representative of the dust that is to be collected. Sometimes samples taken from fall-out in ducts or stacks are all that are available, but these should be treated with extreme care as they are unlikely to be representative of the dust in suspension in the gas stream, particularly in particle size. However, such a sample may be useful for chemical analysis and resistivity measurements.

18.7   DUST SAMPLING FROM THE ATMOSPHERE

Sampling from the atmosphere is carried out for many reasons. Amongst these are to monitor health and explosion hazards, to check on the effectiveness of hoods, enclosures, etc and where appropriate to establish whether required Occupational Exposure Limit levels are being met. Atmospheric sampling devices fall into one of a number of different categories, depending upon the method used to collect or detect the dust:

• filtration;

• inertial techniques, impingement, impaction and sedimentation;

• electrostatic precipitation;

- thermal precipitation;
- light scattering.

Further details on each of these techniques are given in Reference 77. Whilst filtration is by far the most commonly used method of collecting the dust, it is generally unsuitable for collecting particles for subsequent microscopic analysis. Small particles tend to penetrate deeply into the fibres of the filtering media whilst larger particles tend to aggregate on the surface of the media. If particle sizing is required, therefore, one of the many other techniques should be used, such as the cascade impactor.

There are many types of static and personal samplers available. Static samplers can vary from a simple deposit gauge or constant-flow sampler where a certain quantity of air is drawn through a filter to sophisticated electronic equipment which can automatically count and size the individual particles. Inside works, it is becoming increasingly necessary to monitor the air which the worker actually breathes. A static monitoring instrument, however well positioned, does not always correctly assess the concentration of dust in the worker's breathing zone, which will often vary widely as he moves about at varying distances from the main source of dust. In such circumstances, personal samplers are to be preferred. A sampling head is clipped to the worker's lapel which carries a glass-fibre filter through which air is drawn by a small battery-operated pump. By a suitable choice of sampling head, it is possible to divide the dust sample into respirable and non-respirable fractions.

18.8   MEASUREMENT OF DUST PROPERTIES

PARTICLE SIZE ANALYSIS
The size of a particle is that dimension which best characterises its state of sub-division. For a spherical, homogeneous particle, the size is uniquely defined by its diameter. In dust and fume control, airborne particles are usually irregular in shape. Consequently, there are many different definitions of particle size. For irregular particles, various 'equivalent' diameters can be obtained by techniques that measure some size-dependent property such as particle volume, surface area, resistance to motion, etc. If an irregularly-shaped particle is allowed to settle in a fluid, its terminal settling velocity may be compared with the terminal velocity of a sphere of the same density settling under similar conditions. The size of the particle is then equated to the diameter of the sphere (Stokes' diameter). This type of equivalent diameter is useful for characterising airborne

particles in situations where the settling behaviour of suspended solids is being studied. Because the assigned size of particle usually depends on the method of measurement, the particle sizing technique should, where possible, simulate the process under evaluation. In dust collectors, where inertial impaction (collision of a particle with a target such as a liquid droplet) is usually the most important mechanism of particle capture, the size analysis of the dust is often carried out using a cascade impactor which operates on the same principle. The device enables a size analysis to be performed on site by passing a sample of the dust-laden gas over a series of impingement filters. To calibrate the device, use is often made of an 'aerodynamic diameter' (diameter of a sphere of unit density having the same settling velocity as the particle).

Microscopy is the only widely used particle-sizing technique in which particles are observed directly. Because an irregularly-shaped particle has an almost infinite number of linear dimensions, there are a number of different 'diameters' that can be used to describe particle size. Amongst these are Feret's diameter, Martin's diameter, minimum linear diameter, maximum linear diameter, diameter of a circle of equal area and diameter of a circle of equal perimeter.

The different sizing techniques produce quite different results and so it is important to be consistent. When presenting the results of a size analysis, the technique used and the 'diameter' selected to describe particle size should always be stated. Methods of determining particle size are listed in Table 18.1.

TABLE 18.1
Methods for determining particle size

Method	Applicable particle size range (approximate)
Sieving	> 44 μm
Elutriation	5–100 μm
Sedimentation (eg Andreasen pipette)	2–50 μm
Centrifugal sedimentation	0.05–3 μm
Optical microscopy	0.3–100 μm
Electrical resistance counters	0.6–200 μm
Cascade impactor	0.2–8 μm

Information on the wide range of techniques available for particle size analysis is given in Reference 77.

147

OTHER PHYSICAL PROPERTIES
These might include measurement of true density, bulk density, moisture content, particle shape, wetability, dustability, explosibility, hygroscopicity, pyrphoricity, stickiness, abrasiveness, etc. Resistivity of dust samples is normally measured by electrostatic precipitator manufacturers.

CHEMICAL ANALYSIS OF DUSTS
Chemical analysis of dust samples can only be carried out in well equipped laboratories and this service is generally available commercially. Once the general composition of dust is known, it will normally prove quicker, and less expensive, if for further samples the elements of particular interest can be specified to the chemist rather than requesting a total analysis.

18.9   PILOT PLANT TESTING
For particularly difficult problems where full information is not readily available on dust and fume collector performance and where it may be difficult to find equipment operating in similar conditions, most reputable equipment suppliers have facilities to carry out tests with pilot plant collectors.

In certain cases, it is possible to carry out these pilot tests in the laboratory, particularly where the dust or fume problem is essentially associated with atmospheric air and can be readily simulated. However, for all process applications and many ventilation problems, it is necessary to carry out such tests on site, operating the pilot collector continuously, directly on the dust-laden gas stream and at the specific process conditions, temperature and pressure.

Pilot units need to be large enough to effectively simulate the operation of a full scale collector to enable scale-up factors to be confidently used. The operation of pilot units and interpretation of the data obtained require considerable experience and expertise, and such tests are almost invariably performed in association with equipment manufacturers. Pilot testing is also expensive and careful planning of the test programme is essential so that the maximum information concerning the effects of various parameters, such as flowrate, temperature, humidity and process variables, can be obtained as quickly and efficiently as possible.

18.10   PERFORMANCE TESTING
Following the start-up of a dust or fume collection unit it is essential that an assessment is made of the performance of the unit for the following reasons:

- to ensure that the collector performance matches the performance specified by the user in the contractual documents;

- to ensure that there are no other shortcomings in the dust or fume collection system which may affect the collector performance;

- to demonstrate that the process plant is meeting any legal emission limits or requirements of the HM Inspectorate of Pollution or Health and Safety Executive;

- to provide the manufacturer with feedback on his design parameters.

The methods used in performance testing are given in BS 893[80] and for less stringent requirements in BS 3405[85].

In principle, the performance of a collector will be established by simultaneously taking gas samples at the inlet and outlet of the unit, measuring the dust burden and then calculating the dust removal efficiency. The siting of the test positions should be chosen with great care, preferably in a straight section of the duct well away from bends or other areas of turbulence. A straight section of ten times the inside duct diameter is preferable and a preliminary pitot survey will indicate whether a test point is satisfactory. It is advisable to give some consideration to test positions at the design stage to ensure that suitable positions exist and free access is available. Very often, it is sufficient only to sample the outlet stream to establish that the concentration of dust emitted to the atmosphere is within the permitted limit.

Before beginning a performance test, particularly if this is a contractual requirement, the manufacturer will wish to check that the plant is fully operational and also that the gas conditions entering the collector are within the limits specified by the user.

The duration of a test will depend on the stability of the process and the quantity of dust in the gas, but usually 2 to 4 hours is sufficient to give consistent results. Many large industrial concerns will have their own test procedures which must be followed. It is normal to perform at least three tests to confirm their reproducibility.

In addition to the pitot surveys to determine gas flowrates and gas and dust sampling to determine dust burdens, it is advisable to record full details of process operation during the period of testing, including any abnormal occurrences or process upsets, so that these can be related to any deviations in the performance of the dust collector.

Quite often the effluent gas conditions achieved on a new process will differ from those for which the dust collection device was originally designed.

To cover this possibility, manufacturers are frequently asked to supply correction curves which show how the performance of the dust collector will be affected by changes in gas volume, temperature, etc. These correction curves may become part of the contractual guarantee and will be used in evaluating the performance test results.

Performance testing must be performed with considerable care, otherwise the results can be highly inaccurate and misleading. If the user has doubts about his capability of carrying out these tests he should use a contract test team. Most major manufacturers employ their own test team to carry out performance tests, with an observer nominated by the user company.

19. REFERENCES, SUGGESTED FURTHER READING* AND ADDRESSES OF NATIONAL STANDARDS INSTITUTIONS

HEALTH AND SAFETY

1. Abbott, J.A., 1990, *Prevention of fires and explosions in dryers*, 2nd edition, Institution of Chemical Engineers, Rugby, UK.
2. Health and Safety Executive, 1970, *Dust explosion in factories*, Booklet No. 22, HMSO. (Out of print)
3. Health and Safety Executive, 1988, *Control of substances hazardous to health regulations*, Code of Practice 29, HMSO.
4. Health and Safety Executive, *Noise at Work Regulations 1989. Noise Guide No. 1: Legal duties of employers to prevent damage to hearing. Noise Guide No. 2: Legal duties of designers, manufacturers, importers and suppliers to prevent damage to hearing*, HMSO.
5. Health and Safety Executive, *Occupational exposure limits*, Guidance Note EH/40/ (updated annually) HMSO.
6. *Best practical means: general principles and practices*, BPM1/88, 1988, HMSO.
7. Wells, G.L., Seagrove, C.J. and Whiteway, R.M.C., 1976, *Flowsheeting for safety*, Institution of Chemical Engineers, Rugby, UK.
8. Masters, K. 1985, *Spray drying handbook*, 4th edition, Godwin.
9. Health and Safety Executive, 1980, *Flame arresters and explosion reliefs*, Guidance note HS(G) 11, HMSO.
10. *Deflagration venting*, NFPA 68, 1988, National Fire Protection Association (USA).
11. Moore, P.E., 1983, Industrial explosion protection, *IChemE Symposium Series No. 69*, Institution of Chemical Engineers, Rugby, UK, 219–238.
12. *Safety and health at work*, TUC Handbook, 1983, 2nd edition, Trades Union Congress, London.

ENCLOSURES, HOODS AND DUCTING

13. *Foundry ventilation and dust control*, 1955, British Cast Iron Research Association.
14. Flux, J.H., 1976, Progress in secondary fume pollution control in electric arc steel making, *Steel Times Annual Review*.

---

* The Library and Information Service of the Institution of Chemical Engineers in Rugby, UK, offers a worldwide service for the supply of these references.

15. *Industrial ventilation*, 1978, 15th American Conference of Governmental Industrial Hygienists, Committee on Industrial Ventilation, USA.

16. Fletcher, B., 1977, Design of local exhaust ventilation hoods and slots, *Conference on Control of Air Pollution in the Working Environment, Stockholm.* Also HSE SIR 22, 1989, *Reliability of airflow measurements in assessing ventilation systems.*

17. Heriot and Wilkinson, 1979, Laminar flow booths for the control of dust, *Filtration and Separation*, 16 (2): 159–164.

18. Health and Safety Executive, 1988, *Ventilation of the workplace*, EH22 (rev) and/or HS(G)37.

19. Folwell, J., 1978, Design of hoods for dust control systems, *Dust Control, Institution of Chemical Engineers Symposium, Salford.*

20. Crawford, M., 1976, *Air pollution control theory*, McGraw-Hill, Chapter 5.

21. Hemeon, W.C.L., 1963, *Plant and process ventilation*, 2nd edition, The Industrial Press, New York.

PLANT SELECTION

22. Stairmand, C.J., 1970, Selection of gas cleaning equipment, A study of basic concepts, *Filtration and Separation*, 7 (1): 53.

23. Swift, P., 1969, Dust control in industry, *Steam and Heating Engineer*, Vol. 39.

24. Parker, A. (ed), 1977, *Industrial air pollution handbook*, McGraw-Hill.

25. Strauss, W., 1975, *Industrial gas cleaning*, 2nd edition, Pergamon Press.

26. Nonhebel, G., 1972, *Gas purification processes for air pollution control*, 2nd edition, Butterworths, Chapter 12.

27. Heriot, N.R., 1980, A systematic procedure for the control of dust, *Filtration and Separation*, 17 (5): 418–425.

28. Lee, R.W., 1981, Cooling process gas from non-ferrous smelting prior to fabric filtration, *Filtration and Separation*, 18 (6): 536–539.

CYCLONES AND INERTIAL SEPARATORS

29. Koch *et al*, 1977, New design approach boost cyclone efficiency, *Chemical Engineering*, 84 (24): 80–88.

30. Doerschlag, C. and Miczek, G., 1977, How to choose a cyclone dust collector, *Chemical Engineering*, 84 (4): 64–72.

31. Stern, A.C. (ed), 1978, Source control by centrifugal force and gravity, *Air pollution*, Vol. III, McGraw-Hill.

32. Licht, W., 1980, Air pollution control engineering, Marcel Dekker Inc., Chapter 6.

33. Theodore, L. and Buonicore, A.J., 1976, *Industrial air pollution control equipment for particulates*, CRC Press Inc., Chapters 3 and 4.

34. Crawford, M., 1976, *Air pollution control theory*, McGraw-Hill.

WET DEDUSTERS AND DEMISTERS
35. Calvert, S., 1977, How to choose a particulate scrubber, *Chemical Engineering*, 84 (18) :54–68.
36. Calvert, S. *et al*, 1972, *Wet scrubber system study, Vol. 1, Scrubber handbook*, National Technical Information Services (USA), PB 213016.
37. Theodore, L. and Bounicore, A.J., 1976, *Industrial air pollution control equipment for particulates*, CRC Press Inc., Chapter 6.
38. Crawford, M., 1976, *Air pollution control theory*, McGraw-Hill, Chapter 9.
39. Nonhebel, G., 1972, *Gas purification processes for air pollution control*, 2nd edition, Butterworths, Chapter 15.
40. *McIlvaine scrubber manual*, The McIlvaine Company, Illinois.
41. Allen, R.W.K., 1982, Electrostatically augmented wet dedusters, *Filtration and Separation*, 19 (4): 333–340.
42. Muir, D.M. and Miheisi, Y., 1979, Comparison of the performance of a single and two-stage variable-throat venturi scrubber, *Atmospheric Environment*, 13 (8): 1187–1196.
43. Muir, D.M. and Akeredolu, F., Maximising the performance of a multiple-stage variable-throat venturi scrubber, *Atmospheric Environment*.

ELECTROSTATIC PRECIPITATORS
44. Oglesby, S. and Nichols, G.B., 1978, Electrostatic precipitation, Marcel Dekker Inc.
45. Crawford, M., 1976, *Air pollution control theory*, McGraw-Hill, Chapter 8.
46. Theodore, L. and Buonicore, A.J., 1976, *Industrial air pollution control equipment for particulates*, CRC Press Inc., Chapter 6.
47. Licht, W., 1980, Air pollution control engineering, Marcel Dekker Inc., Chapter 7.
48. Bright, A.W., 1979, Fundamentals of dust precipitation in an electrostatic field, *Filtration and Separation*, 16 (3); 284–292.
49. Katz, J., 1978, Maintenance and operation of electrostatic precipitators, *Journal of Air Pollution Control Association*, 28 (9): 868–870.
50. Schneider, G.G. *et al*, 1975, Selecting and specifying electrostatic precipitators, *Chemical Engineering*, 82 (11): 94–108.
51. Feazel, C.E., 1975, *Symposium on electrostatic precipitators for the control of fine particles*, National Technical Information Service (USA), PB 240 440.
52. Potter, E.C., 1976, *Looking differently at electrostatic precipitation technology*, Presented at 69th Annual Meeting of the Air Pollution Control Association, Portland, Oregon.

53. Venditti, F.P. *et al*, 1979, *Symposium on the transfer and utilisation of particulate control technology, Vol. 1 Electrostatic precipitators*, National Technical Information Service (USA), PB 295 226.

FABRIC FILTERS

54. Kraus, M.N., 1979, Baghouses; separating and collecting industrial dusts, *Chemical Engineering*, 86 (8): 94–106.
55. Kraus, M.N., 1979, Baghouses; selecting, specifying and testing industrial dust collectors, *Chemical Engineering*, 86 (9): 133–142.
56. Dennis, R. *et al*, 1975, *Fabric filter cleaning studies*, National Technical Information Service (USA), PB 240 372.
57. Billings, E. *et al*, 1970, *Handbook of fabric filter technology, Vol. 1: Fabric filter systems study*, National Technical Information Service (USA), PB 200 648.
58. Rothwell, E., 1980, Fabric dust filtration: principles and practice, *Filtration and Separation*, 15 (5): 471–474.
59. Theodore, L. and Buonicore, A.J., 1976, *Industrial air pollution control equipment for particulates*, CRC Press Inc., Chapter 7.
60. Crawford, M., 1976, *Air pollution control theory*, McGraw-Hill, Chapter 10.

DEEP BED FILTERS

61. Davies, C.N., 1973, *Air filtration*, Academic Press, London.
62. Batel, W., 1976, *Dust extraction technology: principles, methods, measurement techniques*, Technology Ltd.
63. Dorman, R.G., 1974, *Dust control and air cleaning*, Pergamon Press.

FANS

64. Daly, B.B., 1978, *Woods practical guide to fan engineering*, Woods of Colchester.
65. Osborne, W.C., 1979, *The selection and use of fans*, Design Council Engineering Design Guide 33, Oxford University Press.
66. McFarland, J., 1979, Fans for filtration, *Filtration and Separation*, 116 (2): 144–149.
67. BS 848 *Fans for general purposes*, Part 1 — Methods of testing performance, 1980, Part 2 — Methods of noise testing, 1985.

DISCHARGE OF STACK GASES

68. *The Environmental Protection (Prescribed Processes and Substances) Regulations 1991*, SI No. 472/91, HMSO.

69. Nonhebel, G., 1972, *Gas purification for air pollution control*, 2nd edition, Butterworths.
70. Scorer, R.S., 1978, *Environmental aerodynamics*, Wiley.

ECONOMIC EVALUATION OF PLANT

71. *Dust collector proposal evaluation*, Industrial Gas Cleaning Association, London.
72. Allen, D.H., 1990, Economic evaluation of projects, 3rd edition, Institution of Chemical Engineers, Rugby, UK.

SAMPLING AND TESTING DUSTY GASES

73. Munns, D.D.B.H., 1977, Practical aspects of sampling particulate emissions to the atmosphere, *Filtration and Separation*, 14 (6): 656–663.
74. Hawksley, P.G.W., Badzioch, S. and Blackett, J.H., 1961, *Measurement of solids in flue gases*, BCURA.
75. Stairmand, C.J., 1951, Sampling of dust-laden gases, *Trans IChemE*, 29: 15–44.
76. Nonhebel, G., 1972, *Gas purification processes for air pollution control*, 2nd edition, Butterworths, Chapter 14.
77. Allen, T., 1975, *Particle size measurement*, 2nd edition, Chapman and Hall.
78. Coxon, W.F., 1959, *Flow measurement and control*, Heywood.
79. Hayward, A.T.J., 1979, *Flowmeters*, MacMillan Press.
80. BS 893 *Method for the measurement of the concentration of particulate material in ducts carrying gases*, 1978.
81. BS 1042 *Methods for the measurement of fluid flow in closed circuits*, Part 1 — Pressure differential devices, Section 1.1, 1981, Part 2A Pitot tubes, 1973.
82. BS 1747 *Methods for measurement of air pollution*, 1969–1987.
83. BS 1756 *Methods for the sampling and analysis of flue gases*, 1971–1977.
84. BS 2011 *Basic environmental testing procedures*, 1973–1990.
85. BS 3405 *Method for measurement of particulate emission including grit and dust (simplified method)*, 1938.
86. BS 3406 *Methods for the determination of particle size of powders*, 1986–1988.
87. BS 3928 *Method for sodium flame test for air filters*, 1969.

ADDRESSES OF NATIONAL STANDARDS INSTITUTIONS
British Standards Institution
2 Park Street
London W1A 2BS

ISO
1 rue de Varembe
Case postale 56
1211 Geneva 20
Switzerland

DIN
Beuth Verlag GmbH
Postfach 1145
D-1000 Berlin 30
Germany

Association Francaise de Normalisation
Cedex 7
920 80 Paris — La Defense
France

Nederlands Normalisatie-Instituut
Polakweg 5
PO Box 5810
2280 HV Rijswijk
ZH
Netherlands

Ente Nazionale Italiano di Unificazione
Piazza Armando Diaz 2
1 20123 Milano
Italy

American Society for Testing and Materials (ASTM)
1916 Race Stret
Philadelphia 3
Pa 19103
USA

American National Standards Institute Inc (ANSI)
1430 Broadway
New York
New York 10018
USA

American Conference of Governmental Industrial Hygienists (ACGIH)
PO Box 1937
Cincinnati
Ohio 45021
USA

Standards Council of Canada
Mississauga
Ontario
Canada

Bureau de Normalisation
Quebec City
Quebec
Canada

INDEX